3/23

TOTAL GARBAGE

A Messy Dive into Trash, Waste, and Our World

Illustrated by
JOHN HENDRIX

REBECCA
DONNELLY

HENRY HOLT AND COMPANY
New York

Henry Holt and Company, *Publishers since 1866*

Henry Holt® is a registered trademark of Macmillan Publishing Group, LLC

120 Broadway, New York, NY 10271 • mackids.com

Library of Congress Cataloging-in-Publication Data is available.

ISBN 978-1-250-76038-8

Our books may be purchased in bulk for promotional, educational, or business use. Please
contact your local bookseller or the Macmillan Corporate and Premium Sales Department at
(800) 221-7945 ext. 5442 or by email at MacmillanSpecialMarkets@macmillan.com.

First edition, 2023 / Designed by Veronica Mang

Printed in the United States of America by Lakeside Book Company, Crawfordsville, Indiana

10 9 8 7 6 5 4 3 2 1

For everyone who's ever wondered if
there's such a place as "away"

—*R. D.*

INTRODUCTION

On the second floor of the Manhattan East Sanitation Garage Number 11, over 45,000 objects are on display in a museum most people will never see. Treasures in the Trash is the life's work of Nelson Molina, a retired sanitation worker who spent over thirty years saving things from the landfill while he worked his garbage route in Harlem. You can find Star Wars toys, rows of Pez dispensers, a Christmas corner with a life-size Santa, and even a signed letter from the White House, all pulled from bags and bins waiting to be emptied.

Treasures in the Trash isn't open to the public, but Molina hopes his collection will eventually be moved to a space where "the people can come and see it," he says. "People go to a museum, they see these artifacts.

What I think people ought to see in a museum are what a sanitation worker picks out of the garbage that New Yorkers throw out every single day."

Across the ocean, sanitation workers in Turkey rescued two hundred books from the trash and opened a library that now has over six thousand volumes! The library lends books to schools, prisons, and other programs around the country.

Have you ever heard the saying *A weed is any plant that's in the wrong place?* The same goes for your trash: If you don't want it, and you throw it in the garbage can, it's garbage, right?

Is it still garbage if someone like Nelson Molina finds it and puts it in a museum?

In this book, we're going to learn about trash by asking a few simple questions: What is garbage, where does it come from, where does it go, why do we make so much of it, and how can we do better? We're also going to ask just how bad the trash problem is for ourselves and our planet, and if our garbage has anything to teach us.

The answers to those questions, and to almost every other question about your garbage, get pretty complicated. So put your gloves on and let's take a peek in the trash!

Chapter One

WHAT IS GARBAGE?

Everything in the Treasures in the Trash museum, technically, is trash. In this chapter, we'll take a closer look at garbage throughout history, but keep this idea in mind: How much of our garbage is really worthless? Or does "trash" have different meanings, depending on who's looking at it?

OLD GARBAGE, NEW GARBAGE

Near the shores of the Mediterranean 2,500 years ago, a child played with a homemade doll. The doll's body was made of rags, perhaps scraps left over from sewing clothes, and its head was made of clay. In the present day, across the ocean in North America, a child uses a tablet to play a video game, designing an avatar and choosing clothes for it.

The rag doll ended up in a scrap heap thousands of years ago, and the tablet is probably headed to the trash sooner or later. As long as there have been humans, there has been garbage. But as the doll and the tablet show us, people might have stayed the same throughout history, but our garbage has changed a lot.

What's the difference between trash from a few millennia ago and trash today? For one thing, the kind of stuff we throw away has changed. For most of human history, trash has consisted of natural materials, including food scraps, wood, shell, clay pottery, stone, and bone.

The child who played with the rag doll lived in a world where everything else was made of natural materials too. Plastics and electronics didn't exist 2,500 years ago, and even certain types of metals, like stainless steel, hadn't been invented yet. In the pre-industrial world, things were made by hand, which meant that they were made more slowly and therefore they were considered more valuable. If you had to make everything you used, you'd be more likely to take care of it

all and try to fix something when it broke, especially
if you couldn't just go to the store to buy a new one!

This doesn't mean that families didn't create garbage before the modern era, but they certainly made
a lot less of it. A household might have dug a pit in
the yard or used an old well to hold their garbage. The

garbage pit would have been full of the discards of ordinary life: bones and other food waste, broken pottery, building materials, and broken tools.

Take a look at your kitchen trash can and your recycling bin. What kinds of materials are in there? You might see some natural materials like food scraps, or things made from a combination of natural materials and synthetic chemicals, like printed paper. You probably see a lot of plastic.

If you have a tablet or computer at home, its case is probably made of plastic, and you'll find plastic in its inner workings as well. Think back to the rag doll: If you have a doll today, chances are it's store-bought and made of plastic. You bring it home and remove it from its packaging. Some dolls have even been designed for "unboxing," with a series of plastic packaging to go through: plastic wrap around its plastic container, which contains more plastic packaging for each of its tiny pieces of plastic clothing.

The same goes for most of the things in your house. Try going through just one room and see what you can find that has no plastic in it. It's harder than you might think. Even the fabric on your couch is probably made of plastic!

GETTING RID OF TRASH

Nothing much changed in the world of trash disposal for quite a while from the days of the household trash pit or scrap pile. Ancient Athens experimented with town dumps, ruling that no citizen could dispose of trash within a mile of the city limits. Mayans in the Yucatán Peninsula had central dumps too, although they often disposed of trash outside their homes, as did households all over the world. In medieval Paris, heaps of trash around the city were so enormous they made it hard for French soldiers to see invading armies!

Some European city streets in the Middle Ages had a single gutter running down the center to carry rainwater (and urine) away to nearby waterways. Residents often treated them as garbage heaps, blocking the channels with everything from construction rubble to poop to dead animals.

To prevent trash from piling up, some medieval cities created official dump sites outside city boundaries and charged fines to anyone who was caught dumping their waste in the streets. In Coventry, England, a law stated that all animal dung had to be carted away. Dung-covered streets were such a problem that some cities provided dung carts for citizens to use! Medieval farmers knew the value of all that poop: It was collected and used as manure for local fields.

But things we don't need anymore haven't always been tossed or trashed. Wherever there has been a profit to be made from other people's castoffs, there has been someone willing to try. The "rag-and-bone man" sounds like a character from a scary story, but for centuries, these peddlers would collect scraps and broken objects from American households, offering more useful things like pots, pans, or tools in exchange.

Peddlers could sell scrap metal to manufacturers and things like ashes and cooking grease to soap makers. A few scraps of fabric might not be useful to one family, but in the days before paper was made from wood pulp, a peddler could collect a large quantity of rags and sell them to a papermaker. Peddlers often followed regular routes in rural areas, where a trip to the dry goods store in a wagon over dirt roads could take all day.

Waste pickers were another essential part of the garbage economy. In large American cities after the Civil War, formerly enslaved people and immigrants often did the work of separating different types of waste straight from the trash piles of wealthier people and selling anything of value to junk dealers.

While the wealthy might have thought that their old rags, rubber, wool, bottles, and metals were worthless, the waste pickers understood their value and turned countless piles of castoffs into cash for themselves and their families.

Even children had a role in the business of making money from garbage. In nineteenth-century London, children known as mudlarks scoured the banks of the River Thames at low tide for coal, nails, or anything else they might be able to sell to earn a few pennies. As recently as the mid-twentieth century in Finland, children would comb through waste piles and sell the paper they found to buy candy and toys.

As more goods were made in factories instead of by hand, we started to make more garbage. Suddenly it became easier to buy new things—and easier to throw away old things. The Industrial Revolution also brought more people crowding into cities in search of jobs and new experiences, and where there are lots of people, there is lots and lots of trash.

One common way of handling edible garbage in cities and rural areas was pig-keeping. Some cities in the United States allow people to keep chickens today, but in the 1800s, pigs and horses were a common sight too. At one point, New York City was home to 100,000 pigs, usually owned by members of the working class who let their hogs graze on garbage before slaughtering them for food: literal trash to treasure!

When they weren't inviting pigs to eat their food scraps, New Yorkers and other city dwellers relied on ocean dumping to dispose of their mounting waste. Garbage was loaded onto barges and dumped at sea or in rivers. No one except frustrated city officials liked this solution because the garbage often returned on the tide or showed up in a downstream community.

In the 1890s, Washington D.C.'s practice of dumping its waste into the Potomac River angered the citizens of Alexandria, Virginia, so much that some of them captured a flat-bottomed, garbage-hauling boat and set fire to it. That might seem like an extreme way of dealing with the problem, but imagine the monthly garbage output of D.C. washing up in your town: It included not only thousands of tons of household trash, but hundreds of dead horses and a thousand dead dogs.

As more people crowded into cities from other countries, mainly from Europe, and from rural parts of the United States, household trash became an ever-larger problem. None of the usual solutions were ideal, as they came with their own complications. Pigs might eat trash, but they leave something behind too.

Imagine the stench created by thousands of city-dwelling pigs!

In the mid-nineteenth century, New York City began an experiment with a new technology called waste reduction. This isn't the same reduction you're probably familiar with from "reduce, reuse, recycle." In this case, reduction means to literally reduce the volume of garbage (mainly food scraps and dead animals) by cooking it in enormous vats where the solids condensed and were used to make fertilizer for farms. The greasy liquid that was created in this process was then used to make soap and candles.

The city put all its reduction plants on Barren Island near Brooklyn, and soon a community of mostly immigrant and African American workers and their families grew on the island. Eventually a one-room schoolhouse was built for the island's children, but after the single teacher left, no replacement came for eighteen years.

While the reduction process removed a lot of the city's trash from the public eye, it was still very much in the public nose. As you can imagine, boiling large quantities of dead horses (so many that the waters surrounding Barren Island became known as Dead Horse Bay) created such a foul smell that the wealthy owners of nearby estates complained that it was ruining their health and happiness. Chances are, they weren't very worried about the health or happiness of the people who worked and lived on Barren Island, but they had a good reason to be concerned about the stink. Until the discovery of germs, many people thought disease could be spread by bad smells.

New York also built the country's first municipal incinerator, a controlled facility designed to reduce the volume of trash by burning it down to ash. The ash could then be used as filler in construction but was often dumped in waterways or in open pits just as before.

A cheaper solution was simply to dump trash into open pits on the outskirts of cities, in low-lying areas, or along coasts, but that practice brought all the same problems found by leaving trash on city streets, from the stench to rat infestations. It also added new problems that came from having so much garbage collected in one place. Toxic gas filled the air as masses of garbage decomposed, causing air pollution and even fires. In some places, bears were seen scavenging in open pits.

Today, any of your household trash that doesn't get recycled or reused probably ends up in a modern sanitary landfill, although a few cities run modern incinerators. Sanitary landfills have been the most common way to dispose of our trash in the United States since the first one opened in the 1930s.

MUNICIPAL SOLID WASTE

The official name for all the stuff that we throw away at home, at school, and in our restaurants, stores, and offices is municipal solid waste, or MSW. MSW covers

all the things we toss regularly: paper and cardboard, glass, metals, plastics, yard waste, food scraps,
wood, rubber and leather, textiles (like clothing),
and miscellaneous trash that doesn't easily fit into
any of the other categories. It doesn't include things
like construction debris, industrial waste, hazardous
waste, or electronics. It doesn't count medical waste,
although it does count other types of everyday waste
from hospitals.

The Environmental Protection Agency, or the EPA,
tracks the amount of municipal solid waste American
communities make every year. In 2018, we made an
average of 4.9 pounds of trash per person per day, for
a total of 292 million tons. That's the weight of over
5,600 ships the size of the *Titanic*! The biggest waste
category is paper and cardboard. Next comes food
scraps, then yard waste and plastics, and then metals
like aluminum (soda cans) and steel (soup cans).

The average is 4.9 pounds, so you might have made
less than that, or possibly more. Imagine if you had
to carry it all around with you for a week, as students
in some environmental studies classes do! Our daily
personal trash output has stayed pretty steady for the
past few decades, but our total garbage output keeps
going up. In 2018, the country made 84.1 million tons
more garbage than in 1990. Why? Because today we

have a larger population, which means more people making trash.

The EPA also tracks where we send our trash. In 2018, we recycled 24 percent of our trash and composted 9 percent of it. After incinerating 12 percent and handling some food waste in other ways, the rest, 50 percent, went to the landfill. That's half of our trash in landfills. The biggest category of landfilled waste: food. Food makes up over 24 percent of everything we send to landfills. Plastics is just behind at almost 19 percent. We'll look more closely at how we make all this garbage and exactly what happens on each of these trash pathways later on.

AWAY

Why should we think about what happens to our garbage after it's left our homes and curbside bins? One answer might be that it's never *really* gone.

Imagine a candy bar in a wrapper made of flexible plastic. You eat the candy bar, you throw the wrapper *away*, someone empties the garbage can and takes it *away*, and eventually it just goes *away*. *Away* might mean that you don't have to think about that piece of trash anymore, but it doesn't mean it disappears completely.

We know about people in the past because many of their possessions are still with us today, either in museums or still buried in the ground or in a garbage heap. Maybe, after it had been played with for years and years and its owner outgrew it, that rag doll from the beginning of this chapter was thrown away. But how far away is *away* if the doll was lying in a trash pile, waiting to be rediscovered by an archaeologist more than two thousand years in the future? What if *away* doesn't really exist because, even when we think we've gotten rid of something, it's actually still there?

And what does that mean for us and our garbage today?

WHY SHOULD YOU CARE?

Trash is, by definition, the stuff we don't care about anymore. Even the way we use the words trash and garbage to refer to things we don't like says something about what we value and what we don't. What is valuable and what is trash varies by time and culture too. It might have been common to use rags for doll clothing 2,500 years ago, or even a hundred years ago, but how many of us use rags that way today?

Nothing becomes garbage until someone decides to get rid of it. Why do we decide to get rid of things? Usually, it's for one of a few reasons: It's used up, or broken, or it doesn't fit, or we have no further use for it, or we simply decide we don't want it anymore. Have you ever thrown out something perfectly good just because you decided you were too old for it, or it was out of style, or because you were running out of space?

In this book, we'll explore where our trash comes from and who's responsible for it. But no matter where that responsibility begins, we still have choices to make about the kinds of things we buy or don't buy, and what we do with them when we don't want them anymore. That's one reason you should care about garbage: It's personal.

Another reason is that garbage isn't just ugly, like a street covered in plastic bottles and bags. It's actually bad for the environment in a lot of ways. The nineteenth-century doctors who thought garbage caused disease might have been wrong in some ways, but we'll see how your garbage plays a bigger part in the health of our planet—and you—than you might have realized.

Chapter Two

WHERE DOES OUR GARBAGE COME FROM?

Garbage comes from everywhere. If garbage is just the name we give to things we don't want anymore, almost everything you and your family own is future garbage. You might want and need your possessions now, but what happens next week or next year? As soon as you put something into a garbage can or a recycling bin, whether it's a single-use plastic water bottle or a scruffy old jacket, it's garbage.

In this chapter, we'll check out different categories of trash, but first, let's get our feet wet in the waste stream.

DOWNSTREAM AND UPSTREAM
IN THE WASTE STREAM

We call the flow of trash from our homes and businesses "the waste stream." Sounds almost pretty,
doesn't it? No? The waste stream might not be good
for fishing, but it's a useful way to think about how
garbage is produced and where it goes. Every product
you use and every scrap of food you eat has "downstream" and "upstream" waste issues.

Think about the last time you got a drink from
the soda fountain at a fast-food restaurant. The
upstream waste issues are all the things that had to
happen to get that soda into your hands. The downstream issues include everything that happens to
it when you're done drinking. Every product goes
through a life cycle, which shows how it's made,
how it gets to consumers, and what happens when
it's used up.

We'll start with the cup. A factory had to get raw
materials to make the cup, the lid, and the straw. This
is called resource extraction. The paper in your cup
started out on a tree farm. Heavy machinery had to
come in and cut the trees. Even if your cup was made
with some recycled paper, most of it will be new pulp.

The logging machinery needed gas to run its engines, which also had to be extracted. (We won't even get into the materials and energy needed to make the logging machinery!) Unless you're using an eco-friendly all-paper cup, your cup has a plastic liner to prevent leaking. Plastic comes from oil or natural gas—that's more resource extraction!

Turning raw materials into finished products is the production phase. More machinery and energy get used to turn the trees into pulp. Bleaching chemicals are added, and the pulp is made into plastic-lined paper. The paper might be printed with a design or logo. Now the paper is ready to be molded into a cup, and the bottom is attached by heat sealing, which melts some of the plastic to form a seal.

Once the cup is ready, it has to get to a store or restaurant so you can buy it. This is called distribution. In the U.S., most distribution happens in large trucks, which also require materials and energy.

The straw and lid have to go through resource extraction, production, and distribution too. Looking at all these upstream and downstream costs is a product's life cycle assessment. A measurement called a carbon dioxide equivalent (CO_2e) is often used in life cycle assessments to see what kind of impact different products have on the climate. Using ten paper cups with lids has the same CO_2e as driving a car one mile.

All this, and we haven't even gotten to the drink yet! Your soda probably began its life in a cornfield. The corn had to be harvested and sent to a factory to be made into high-fructose corn syrup, the most popular sweetener in soda. The artificial flavors and colors were created in a lab to be added to the corn syrup. The syrup was sent to the fast-food restaurant. And don't forget, all the soda fountain equipment that does the magic of turning syrup into soda by adding water and carbonation has a life cycle of its own.

This is a quick version of the upstream part of your soda's life cycle. At every step along the way, there are opportunities to create waste. For every garbage can you take to the curb on trash day, seventy garbage cans of waste were created upstream to make all the stuff you threw away. Is the waste stream starting to look more like a flood?

Drinking that soda is the consumption stage, and it's probably the shortest part of its life cycle. Think of all the resources that were used to make a drink that lasted you ten minutes! When you've finished your drink and you're wondering what to do with the cup, straw, and lid, there are more opportunities for waste. These are the downstream issues, or the disposal stage.

PAPER AND PAPERBOARD

In Chapter 1, we took a quick look at municipal solid waste, or MSW. Now, let's dig a little deeper.

If you made a list of all the things your family throws out in a week, including everything you put in the recycling or compost, the biggest category by weight would probably be paper and paperboard (that's MSW-speak for cardboard), followed by food waste, plastics, yard trimmings, and metals.

The paper and paperboard category includes newspaper, printer paper, corrugated cardboard and the thin cardboard that's used in packaging, milk cartons, paper bags, and things like tissues and paper towels. Every shoebox, every coloring book, every old worksheet you toss out makes up part of the 67 million tons of paper and paperboard we use in a year. If all our paper products were made from new fiber, that would be like cutting down over 350 million trees!

FOOD WASTE

Americans waste more food than any other nation. In 2018, we threw out 63 million tons of food. If an average elephant eats 300 pounds of food in a day, over a million elephants could munch for a year on all that waste! (Whether or not they'd want to is a different question.)

That huge number includes just about every scrap of food waste besides what we put in our backyard compost piles, which doesn't get counted in MSW. It comes from our homes, schools, restaurants, grocery

stores, and places like hotels and sports arenas. We threw out almost 25 million tons at home, and another 17 million was wasted in restaurants.

A study in 2019 showed that on average, students waste 39.2 pounds of food per person per year and 28.7 cartons of milk at school. Together, all U.S. schools that participate in the National School Lunch Program make an estimated 530,000 tons of food waste per year. Because most of that goes to landfills, which produce greenhouse gases, that's like putting an extra 46,100 cars on the road!

Some food waste is hard to cut back on, like carrot tops and apple cores, things you're not likely to save and eat later. A lot of it is food that could have been eaten but was tossed instead, like the fries you didn't finish, or that weird jar of sauce in the back of the fridge that grew mold.

Why do we waste food at home? Pick up a gallon of milk, a can of soup, or a box of crackers. Somewhere on the package you'll find a date. It might read "sell by" before it, or "use by," or "best if used by." Often, we trash food because it's near that date, even if the food inside is still perfectly edible. When that happens, not only do we waste the food and all the water and energy that went into growing and processing it, we waste the packaging too.

Of course, plenty of food really does go bad before we can use it. Often, that's because we've bought more than we can use, or we bought something new that we didn't like. Sometimes we buy fresh vegetables and meat, but we don't have the time to cook them before they get moldy or start to spoil.

Some of our food gets wasted even before we have the chance to bring it home. It might seem strange that a grocery store that makes money from selling food would throw food away, but that happens every day, creating over 8.5 million tons of food waste each year. Often, it's for the same reasons we throw away food at home. A grocery store will offer a discount on food that's getting close to the sell-by date, but if the food isn't sold in time, it's headed for the dumpster.

If you walk around the produce section of a large supermarket, you'll see an enormous variety of fruits and vegetables stacked attractively. Grocery stores can refuse to buy any produce from their suppliers that looks less than perfect. Some of this "ugly" produce might not necessarily be wasted—blemished apples, for example, can be made into applesauce, apple pie, or something else where looks aren't as important. But even veggies that are considered too small or oddly shaped can be rejected.

Remember how the waste stream shows us the long journey stuff takes before it becomes trash? Going further upstream, we see that food also gets wasted before it gets to the store. A lot of the crops rejected by supermarkets have nowhere else to go and end up rotting in the field. Food wasted from the farm, food storage facility, or food processing factory is called food loss. Add food loss and food waste together, and it's estimated that about 30 to 40 percent of the food we produce is never eaten.

PLASTICS

Although all types of waste can cause environmental problems, plastics might be the worst villain in the story of trash.

The United States tossed more than 35 million tons of plastic waste in 2018. That's the equivalent of about four and a half trillion small rubber duckies (or one really big rubber ducky). Think about how many things you use in a day that are made of plastic. If you pour yourself a bowl of cereal in the morning, does it come in a plastic bag to keep it fresh? Do you pour it into a plastic bowl? Does your milk come in a plastic gallon jug? Your toothbrush is probably plastic, and your backpack is likely made of a plastics-based waterproof fabric. Same for your insulated lunch bag and the sandwich bag inside, or even the reusable sandwich container. Plastic can take so many forms, it's sometimes hard to recognize. That supersoft microfleece jacket and hat that keep you warm? Also plastic.

Plastic is lightweight, compared to things like steel. It's cheap—at least for now. As we saw in the life cycle of the disposable cup, most plastic comes

from oil and natural gas—the same fossil fuels that heat your home, power your car, and provide electricity. When oil is cheap, plastic is cheap. But new plastic is a nonrenewable resource. The fossil fuels that we use to make it formed underground over millions of years from decomposing prehistoric plants and animals. When we take them out of the ground to use as fuel or make into cute bath toys, we have no way to make more fossil fuels. If you're wondering why something that took millions of years to make is so cheap, you're not alone, but that's a topic for another book.

Millions of years is a lot of time to spend for a plastic bag that gets thrown away after a single use!

TEXTILES

Have you ever left your jacket on the floor and been told to pick it up because your home isn't a garbage dump? Plenty of people send their unwanted clothes to actual dumps, along with other textiles like sheets, towels, and carpets. Altogether, Americans trashed 17 million tons of textiles in 2018. Almost 13 million tons of that was clothing and footwear—that's the weight of around 6.5 billion pairs of sneakers. And although some textiles can be recycled, most end up in the landfill. We recycled about 2.5 million tons, or 15 percent, of our textiles in 2018.

People throw out clothes for all kinds of reasons. Things that are too small or too stained or too out-of-date often get dumped. Maybe it takes too much time to send it all to a thrift store or have a yard sale or pass it on to someone else. These days we throw out way more clothes than we used to, over 800 percent more than your grandparents did in 1960. Part of the reason is something called fast fashion.

Fast fashion literally speeds up the whole process of designing and selling clothes. Some fast fashion companies put out up to twenty thousand new styles of clothing in a year. The price is cheap, but so is the quality. Cheap clothes wear out more quickly. On average, we're not wearing our clothes as many times as we used to before getting rid of them. We ditch about half of the fast fashion pieces we buy in less than a year!

ELECTRONICS

Although it's counted separately from MSW, let's take a quick look at another type of trash that you probably make. Electronic waste, or e-waste, is a pretty small percentage of the total trash we make every year, but it counts for about 70 percent of our hazardous waste. In 2018, we got rid of 2.7 million tons of e-waste—that's computers, game systems, phones, tablets, TVs, cameras, DVD players, and any toys that have microchips in them to make them talk, sing, light up, and so on.

Altogether, the amount of e-waste the world makes in a year weighs the same as 125,000 Boeing 747 jumbo jets!

Don't forget all the plastics used in your electronics too, from your laptop or game system case, the body of that singing toy, and even the insulation that covers the copper wires inside.

SCHOOL WASTE

What kind of garbage comes from schools? We could make a joke about homework here, but let's not make your teacher mad.

We don't have national figures for all the waste created in American schools, but studies in Minnesota and South Carolina measured the waste created per student at anywhere from half a pound to a pound per day. If you take the average of those numbers, three-quarters of a pound per day, and multiply it by the number of students that were expected to enroll in pre-K through twelfth grade in 2020, that's 42.3 million pounds of waste for the year, or 23,500 pounds per school day.

Food is the most common type of school waste, but school trash cans are loaded with other treasures too, including school supplies like pencils and glue sticks, food packaging and single-use items like foam trays and plastic utensils, paper, batteries, and decorations. About half of this waste could be recycled, composted, reused, or not produced at all.

So, where does our garbage come from? *Everywhere* still seems like a pretty good answer, doesn't it? To be more specific, our garbage comes from home, school, stores, restaurants, and all the other places we buy and use things. Most of our garbage is manufactured, and a lot of it could be avoided. To help us see how to make less trash, we need to know why we make so much to start with.

Chapter Three

WHY DO WE MAKE SO MUCH GARBAGE?

We make a lot of trash.

In fact, we make almost twice as much trash per person today as we did one hundred years ago. What has also changed is the kind of stuff we throw out. We've seen that, until the twentieth century, most garbage came from natural materials like bones and food scraps. In 1920, up to 75 percent of household garbage was ash from fireplaces and stoves. Today, over half of our MSW is manufactured materials.

Remember that soda from Chapter 2? The cup, straw, and lid are all examples of single-use items, things that are made to be used once and then thrown away. Even the few single-use items that get recycled are only used once before they take a dive into the waste stream.

A few generations ago, the idea of using something once and throwing it away would have seemed unbelievably wasteful. These days, we do it all the time. The change from natural to manufactured materials and from reusing things to throwing them away was no accident, and it helps us see why we make so much garbage.

THROWAWAY LIVING

In 1960, Americans tossed 270,000 tons of paper plates and cups. By 2018, that number was up to over 1.4 million tons. That's more than five times as much! We didn't make enough plastic plate and cup waste to bother tracking it in 1960, but by 2018, we were throwing out over a million tons of that too. The story of how that change happened is, in a way, the story of America after World War II.

A 1955 article in *Life* magazine titled "Throwaway Living" showed a surprising photograph of a family standing under a shower of disposable goods: plates, napkins, utensils, pie tins, straws, bags, and more.

That picture is a sign of a change that was happening in America at the time.

In 1955, World War II had been over for ten years. During the war, Americans were told that saving and reusing was patriotic because the military needed all the raw materials and scrap it could get for war manufacturing. A popular slogan was "Use it Up, Wear it Out, Make it Do, or Do Without!" Reusing and repairing were also standard before the war, as we've seen. But that was about to change.

After the war, messages about patriotism sounded very different. Now, Americans were told to spend money to keep the economy strong. Wartime manufacturing had built factories and created jobs across the country, and all those factories wanted to stay in business producing things Americans could use in peacetime, like cars, appliances, and televisions. Americans were told that reusing and repairing things they already owned was no longer patriotic. Instead, buying new was a way to keep people employed in good manufacturing jobs. Although most of those jobs went to white men returning from the war, factory jobs were important to people of color too, and drew many Black families away from southern states in the Great Migration to work in northern industries.

There was another reason this new consumer lifestyle became a symbol of patriotism after World War II. From 1947 to 1991, America and the Soviet Union, led by Russia, engaged in the Cold War. The Cold War led to some true wars in Korea, Vietnam, and other places, but it was also a war of ideas: American capitalist democracy against Soviet communist dictatorships. Communism, which began with the Russian Revolution in 1917, was an economic system that put property and industry in the hands of the government instead of private companies. Americans' gleaming new cars, big houses, and huge department stores, all privately owned, were meant to show that the American way of life was more extravagant, and therefore better than communism.

USE IT ONCE, TOSS IT OUT!

Industry's best sales pitch for wastefulness wasn't to tell people they were fighting communism or even to remind them about American jobs. The point of "Throwaway Living" and similar articles and advertisements was to convince families that buying single-use, disposable items would give them more leisure time and make their lives easier. Many of these ads were aimed at housewives whose days were full of repetitive tasks like washing dishes and changing diapers. The disposable diaper, which hit the market in 1961, might have seemed like a miracle to someone who was used to emptying and cleaning cloth diapers!

It took some time and effort for companies to convince Americans that these new products were only meant to be used once. An American in 1950 might have used disposable paper cups in a public place like a school or train station, where they had been introduced as a way to stop the spread of germs, but using disposable products at home wasn't common. Shoppers expected things to last. If you were used to keeping the same set of dishes for generations, throwing things away didn't come naturally.

Even packaging used to be designed to keep. In the days when many families bought flour in large cloth sacks, flour companies designed the fabric to be reusable. The "flour sack dress" wasn't a cheap or ugly piece of clothing. It might have been decorated with flowers, plaid, or paisley in a variety of bright colors. Jars and

tins were saved and reused. Some tobacco tins were even designed with handles and decorations so they could have a second life as children's lunch boxes.

But all that advertising worked, and soon the American way of life became tied to convenience and consumerism. At one point, a fashion company even offered disposable paper dresses for sale!

PLANNED OBSOLESCENCE

Convincing us to use disposable products was one way of getting consumers to spend more money and keep the assembly lines moving. But manufacturers had other tricks up their sleeves. The easiest way to make sure that consumers kept consuming was to put out lots of products quickly. Assembly lines and mass production in factories made this possible. Many mass-produced goods were made cheaply, and that showed in their quality and life span. Cheap goods often break more easily and can be hard, if not impossible, to repair. If they have to be replaced instead—*cha-ching!*

Styles began to change more quickly too. Automobile companies offered new body styles every year, hoping that even if there was no good reason to replace an older car, drivers would want to buy new cars every few years to keep up with their neighbors. This kind of social pressure can make us feel ashamed of having old cars, worn clothes, and other not-so-new belongings. We don't want people to make fun of us or judge us, and buying the "right" things can make us feel like we fit in. Twentieth-century advertisers began to take advantage of that feeling to get us to buy, buy, buy.

Technological changes also helped shoppers open their wallets. It's the same thing that happens with cell phones today. Once most of us have cell phones, how can a cell phone company keep making money? By adding new features to new models: a better camera, different colors, more memory, or fun software. Your old phone might work perfectly well, but you'll feel tempted to upgrade anyway because having an older phone somehow feels like getting cheated.

These are examples of something called planned obsolescence. Instead of giving us the most value for our money by making things that will last for decades, manufacturers design products that will quickly break, go out of style, or seem boring and outdated.

THE PLASTIC MIRACLE?

The war also made a once-rare material incredibly popular: plastic. At the beginning of the war, only a few common household items were made from plastic. Celluloid, an early type of plastic, was used to make a few things like combs and camera film. When new types of plastic were developed, war manufacturers saw that they could be used as lighter, cheaper replacements for various other materials. Plastic isn't just one substance; it's a type of chemical polymer that can be adjusted to create hard, soft, rigid, or flexible materials, from cockpit windows to waterproof tents.

After the war, plastics manufacturers wanted to keep all those oil refineries and factories busy, so they began to think of new ways to use plastic in ordinary life. Many plastic products were injection-molded. That means liquid plastic was used to fill pre-made molds. The typical plastic patio chair is made from an injection mold, making it cheap and fast to produce. But because it's made from a single piece of plastic, it's also basically impossible to repair, so it quickly becomes garbage.

PACKAGING

Mass production gave us more packaging too. Packaging makes life pretty convenient. Grocery shopping is much faster when you can grab bottles, cans, and boxes from the shelves. Companies like packaging because it makes things easy to transport, and putting brand logos on packaging gives the company built-in advertising. They're betting that you're more likely to buy things from a brand you trust.

Packaging materials are mostly paper, plastic, metals, and glass. Together, packaging of all kinds makes up a little over a quarter of our total MSW—that's 73 million tons!

We talked about food waste in Chapter 2. Often, wasted food means wasted packaging too. Some of that packaging can be recycled, but most of it can't. Pay attention to packaging the next time you go to the store. What kinds of packaging do you see? How many things have packaging that they don't need?

We even "package" our trash—in 2018, the weight of all the plastic trash bags we used was over a million tons!

PANDEMIC GARBAGE

The coronavirus pandemic that began in 2020 had terrible consequences for human health, the global economy, children's education, and many parts of our daily lives. This includes our garbage, and, as we've seen, waste has a way of causing problems downstream.

As Americans spent more time at home, our shopping

habits changed. We were ordering more things online than before, which meant more boxes and other shipping materials. We used more delivery services, which meant using fewer reusable bags and containers. We ate more takeout, using more disposable containers and silverware.

Plastics companies jumped at the chance to tell shoppers that single-use plastics would help stop the spread of the virus. Many states that had banned single-use plastic bags in grocery stores stopped enforcing those bans. Some bulk stores stopped allowing shoppers to refill their own containers.

A type of litter that was pretty rare before 2020 started showing up everywhere: disposable personal protective equipment, or PPE. Did you see blue surgical masks and gloves covering the streets of your town during the coronavirus pandemic? When masks became a symbol of the times, so did PPE in the gutters and on the sidewalks.

The pandemic shows us how quickly our habits can change. Good news, though—this means that we can also change them for the better.

REPAIR

When was the last time you made your bedsheets last longer by cutting them down the middle, reversing them, and sewing them back together with the worn part on the edge and the stronger part in the middle? Never? What about darning a hole in the toe of a sock? If you don't know what darning is, it just shows how these old-fashioned practices have disappeared from our lives.

Why don't we fix more of our household objects? It's not just that a ten-pack of new socks from a big box store is so cheap. When it became easier to replace things than to fix them, we lost many of those skills. It takes time to repair things—time to learn how to do it and time to do the work. We might feel that there are other things we'd rather be doing. Throwing out one pair of socks doesn't make *that* much difference in our trash bins, so why not?

Even if you have the skill, the time, and the parts to repair an object you own, there's one more thing to consider. Companies are in the business of selling you new items, not helping you fix the ones you have. This is particularly true for digital devices. If you try to fix a cell phone yourself, you might end up voiding the warranty, which means that the manufacturer won't pay to fix or replace it when something else goes wrong. Device manufacturers don't usually make repair manuals for consumers either. Your only option is to send it back to the company to fix, and sometimes that can be as expensive as buying a new phone.

WHO WASTES MORE?

It's important to say that not everyone consumes things equally. High-income countries make two to three times as much waste per person as low-income countries, and more of that waste comes from packaging and other non-biodegradable items. In low-income countries, organic waste makes up a bigger share of all garbage. High-income countries tend to throw out more stuff that could be recycled too, even though they're also more likely to have formal recycling programs.

Economic inequality can also create waste. We often see "buy in bulk" or "buy the biggest package you can afford" as a suggestion for making less trash. While this is good advice for people who can afford it, the reality is that many people can only afford to buy food and personal care items in smaller containers, even if it means spending more money and creating more garbage over time.

Residents in Flint, Michigan, had to rely on bottled water for years when their water system was contaminated with lead. Telling people to stop buying single-use bottles doesn't help if their tap water isn't safe to drink. More than half of Flint's residents are Black and nearly 40 percent live under the federal poverty line. There's a term for this issue: environmental racism.

America isn't the only country where we can see how poverty creates waste. In many low-income countries, personal care products are sold in plastic sachets, similar to the ketchup packets you find at fast food restaurants. The sachets hold enough of a product for one use (say, one shower) and they're cheap. For people who earn just enough money to live on each day, a large bottle of shampoo that would last for a couple of months might be too expensive. But the flexible plastic of sachets can't be recycled at all and often ends up as litter.

WHOSE FAULT IS IT?

You might look at a photograph of a shoreline covered in litter, or a massive modern landfill, and feel a little guilty. After all, it's the things we buy and use that end up as trash in one place or another. If you don't pick up a plastic sachet of ketchup with your order of chicken nuggets, there's no way it can end up in a whale's stomach. As a consumer, are you responsible for what happens to the things you buy and use when you're done with them?

We've looked at how some of these choices are made for us. We don't make all that trash on our own. Manufacturers and advertisers decide what kind of packaging to use and what materials should be used to make it. Companies also decide if their products will be long-lasting and how easy they'll be to repair. If consumers can't control how things are made, how can they be responsible for what happens to them when they get trashed?

We've also seen how these choices often come down to money. Simply put, if you have to buy new things, someone makes money. The company that made the product, the advertisers, the store that sold it—even waste haulers make money, because when you throw things away, it keeps the trash companies in business.

Obviously, it's good to have manufacturers and stores and someone to handle waste. And it's nice to have convenience and new things from time to time. But what if manufacturers were less focused on making that process as fast as it can be and more focused on what it meant for the environment? What if consumers had better options? We'll look at some ways to make that happen. But now, let's see where our trash goes.

Chapter Four

WHERE DOES OUR GARBAGE GO?

You probably don't think too hard about the final stage of your trash's life cycle: disposal. That's what happens when your garbage goes away. But remember—there's no such place as *away*.

Your trash won't just disappear. It might get picked up at the curb by a trash hauler or you might take it to your town's transfer station yourself, but however you get rid of your trash, you probably never see where it ends up. A typical garbage truck can pick up waste from eight hundred homes before it's full. So where does it go after that?

The main ways we deal with trash are landfilling, recycling, composting, and incineration. We can also get rid of things we don't want by donating them to secondhand stores or programs that take donated items. But since we're talking trash, let's follow the garbage truck.

LANDFILLS

In 2018, we sent half our garbage to landfills. That figure includes a lot of stuff that could have been recycled or composted instead. New York City estimates that 68 percent of its household trash could be recycled or composted, but in 2017, 80 percent was sent to the landfill!

There have always been open dumps, but modern sanitary landfills have only existed since 1934. For a few reasons, they've become America's favorite way to deal with trash ever since. They're cheap, compared to other technologies. Landfilling all our trash is easy, much easier than separating different kinds of materials and figuring out what to do with each type. And for many of us, they're far away, so after we toss our trash, it's gone.

In the late nineteenth and early twentieth centuries, cities started looking at waste and waste removal as a public health issue. Remember that scientists thought that garbage itself spread sickness through the air, and although that wasn't exactly true, cleaning it up made city streets easier to walk on, safer, and less smelly.

City departments were set up to clean streets and pick up trash from homes and businesses. Sanitation workers in New York City known as White Wings dressed in white uniforms to emphasize their role in keeping the city sanitary.

As we saw earlier, it was common to dump trash in waterways, burn it, or send it to be boiled down in reduction plants. In cities from New Orleans to Seattle, it was also used to fill in marshes and wetlands.

The grassy parks and building sites created from garbage masked a serious problem, though: Decomposing garbage creates gas, and gas catches fire. In Oakland, California, a section of filled waterfront caught fire in the late 1930s, and the thousands of rats that made the dump their home were forced to flee to other parts of the city. In Dayton, Ohio, dumps created such a bad pest problem that insects were reported to be swarming houses, eating food that was left out, and even chowing down on the wallpaper.

Rikers Island, New York, was a dump before it became a notorious prison. So much garbage was brought there—using convict labor—that it grew to twice its original size and boasted trash towers 140 feet tall. The rat population exploded as these adaptable rodents took advantage of the buffet. To control the rats, officials brought in dogs. Soon over a hundred dogs lived on the island, feeding only on the rats that fed themselves on garbage.

Open dumps and filled-in wetlands were cheap, but the problems of fire and pestilence led to a demand for better solutions. Simply covering trash with dirt didn't solve much. A mix of methane and carbon dioxide gases was still created, and animals could easily dig down to the tasty trash beneath the dirt. As rain percolates through trash piles, it picks up hazardous chemicals. This toxic liquid, known as leachate, can then poison soil, groundwater, and waterways.

The sanitary landfill aimed to fix those problems by containing trash in cells and compacting it to squeeze out the oxygen needed for trash to break down. A huge machine called a compactor can crush 1,200 pounds of garbage into one cubic yard of space! The smashed trash is then covered with layers of soil, clay, sand, and gravel. In a modern landfill, a five-foot layer of these materials sits on top of a layer of compacted solid waste.

Sanitary landfills are so good at compacting trash that rats have a hard time digging into the layer of solid waste to retrieve anything that might be edible. In fact, compactors have also been known to compact any rodents that get in the way!

Keeping rats out of landfills has its advantages, but the highly mechanized system that starts with curbside trash collection and ends up at a modern landfill makes scavenging impossible for humans too. In the old open dump system, we saw that waste pickers could find objects that could be reused or recycled and save them from the landfill while also making a little money. Not only are modern landfills often isolated and hard to get into, but the compactors also crush things that might once have had value, making landfills little more than trash graveyards. In fact, sanitary landfills are also known as "dry tomb" landfills. Creepy!

And like a graveyard in a horror movie, scary things can come out of even the best-managed sites. Landfills are designed to have an active life of thirty to fifty years. When they're full, they get another layer of clay and soil and a plastic liner on top to prevent oxygen and water from getting in. The owner of the landfill has to keep an eye on the methane and leachate for another thirty years.

That might sound good, but landfills produce methane for around a hundred years. And no sanitary landfill has reached its hundredth birthday yet, so we don't really know what happens after that. We don't even know exactly how many landfills there are in the U.S. because the EPA doesn't keep track!

A few waste management companies in the U.S. are experimenting with a new technology called the bioreactor landfill. Unlike dry tomb landfills, which are built to keep as much liquid as possible out and prevent garbage from breaking down, bioreactors use liquids like landfill leachate or wastewater to help trash decompose faster. When garbage decomposes, it takes up less space, and it makes more landfill gas that can be captured for energy.

One more fun fact about how well sanitary landfills preserve trash: A team from the University of Arizona found guacamole in a landfill, and by looking at the dates on other things nearby, they figured out that the guacamole was twenty-five years old!

RECYCLING

A little less than a quarter of our total trash was recycled in 2018. You probably know that recycled materials are turned into new things. For example, aluminum cans get baled together, melted in a huge furnace, and turned into large blocks called ingots, some as tall as thirty-two feet. These are sent to a factory to be rolled out into sheets and made into new cans. This process can happen again and again, practically forever.

Aluminum recycling is efficient and profitable because aluminum is stable. Its quality stays the same even after dozens of trips through the furnace. Also, mining bauxite (the ore aluminum comes from) is an expensive process. It's much cheaper to make new soda cans from recycled aluminum, and it saves 95 percent of the energy it would take to mine new bauxite. Unfortunately, not everything in the world of recycling is that straightforward or that efficient.

Unlike aluminum, recycled paper usually has to be mixed with new wood pulp because each time paper is recycled, its fibers get shorter, making the paper weaker. Paper can only be recycled four to six times before the quality deteriorates too much. It still saves landfill space and energy—more than 60 percent over manufacturing new paper.

Paper has the best recycling rate of any material in our MSW. In 2018, we recycled over 68 percent of the paper and paperboard we produced! But we're still sending almost 32 percent of it to be landfilled or incinerated.

Compare that to our dismal plastics recycling rate. In 2018, we only recycled 8.7 percent of all our plastics. To picture the difference, imagine you have a box of thirty cookies. To waste as many cookies as we waste in paper and paperboard, you'd have to eat twenty and a half of those cookies and throw the rest away. To waste as many cookies as we waste in plastics, you'd only be able to eat two and a half cookies!

Some types of plastic have a higher recycling rate than others. Twenty-five percent of the 70 billion plastic water bottles sold in the U.S. in 2018 were recycled. But of all the plastic ever produced, only about 9 percent has been recycled. As we saw in Chapter 3, plastic is everywhere, but not all plastic is truly recyclable. Why? The answer, again, is money.

The recycling rate for plastic is so low because, of all the different kinds of plastic out there, only some can be profitably recycled. Recycling is a business, and in order for recycled plastic to make a profit, someone has to buy it. New plastic is often cheaper and easier to make, so if no one wants to buy your old plastic waste to turn it into something else, it's going to the landfill or the incinerator. And that's exactly where most of it ends up.

Let's take a trip to the typical curbside recycling bin. Most of us have what's called "single-stream" recycling programs, meaning that you dump all your recyclables, from newspapers to soup cans, in one bin and let someone else figure out how to separate everything. Separation happens at a place called a materials recovery facility, or MRF. The idea is that more people will recycle if it's easier to do—that's a good thing. But in practice, it hasn't worked very well.

Curbside recycling programs started out by asking people to separate some of their recyclable items, often putting paper and cardboard in one bin and cans and bottles in another. Any sauce, soup, or soda had to be rinsed out. Then, in the early 1990s, something changed in the world of recycling. China started offering to import huge quantities of the world's waste plastic and paper as it started to become more industrialized and its factories needed more raw materials.

U.S. recyclers developed single-stream programs in order to get more and more material to send to China. Unfortunately, much of it was contaminated and useless. In single-stream recycling, paper can get ruined by liquids and food waste as they mix in the same bin. If customers don't understand the rules of plastics recycling, unrecyclable plastics get mixed in with recyclable plastics. Plastic bags, including shopping bags and trash bags, and plastic wrap are a real problem in curbside recycling because the thin film clogs the machinery at the MRF.

Because these programs encouraged people to toss everything in one bin and not worry about it, weird, unrecyclable junk ended up in the recycling stream, like lawn furniture and plastic storage totes. Putting something into a recycling container and hoping it will get recycled is called "wish-cycling." As in, wishful thinking. The reality is that mixing recyclable and nonrecyclable items can ruin an entire load, and a contaminated load of recycling is just trash.

Partly because the quality of imported waste was so low, China announced that it would no longer accept plastic waste exports from other countries. That left U.S. recyclers scrambling to figure out what to do with all the plastic Americans were putting in their curbside bins. It could all be separated, but if no one wanted to buy it, it wasn't going anywhere.

In 2017, the last year China accepted plastic waste from other countries, the U.S. sent over a million tons. Sending plastic waste abroad counted as "recycling" because that's what we thought was happening. The dirty secret is that we have no idea if our waste was being recycled, burned, or buried. So even our very low figures for recycling plastic waste are probably too high.

We haven't stopped sending plastic waste overseas, though. It's just going to other countries, mainly in Asia and Africa. In May 2019, we sent over 71,000 tons of waste to fifty-eight different countries. Some of it is recycled and used in manufacturing, but much of it is simply burned or left in the environment. In the past, researchers have blamed countries in Asia for most of the world's marine plastic pollution, but that doesn't tell the whole story.

The United States makes the most plastic waste of any country on Earth, and we export a lot of that to countries that aren't equipped to handle it properly. It's estimated that more than 80 percent of the plastic waste in some of these countries is mismanaged. Waste pickers in these countries provide important services, picking out materials that can be recycled, but when so much of the plastic we create isn't recyclable, they're left with mountains of our waste to deal with.

DOWNCYCLING, NOT RECYCLING

Some types of waste can only be made into something new one time. That's called downcycling. Many textiles that are too worn out to be donated to a secondhand store can be "recycled," but textile recycling isn't an endless loop like aluminum recycling. Fabrics can be washed and cut into wiper rags for use in places like factories and automotive shops, or chopped and made into the padding that goes underneath carpets. Once your old T-shirt gets a second life under a carpet, however, that's the end of the line. It won't be recycled again after that. Downcycling is a stop along the way to the landfill.

Like paper, the quality of plastics goes down as they're recycled, so they can't be recycled indefinitely. Only one tenth of your plastic waste will get recycled more than once. Although some plastic bottles will be recycled into new bottles, many will be downcycled into microfleece clothing, benches, or decking material, and when those things are worn out, they're tossed out.

Remember that money drives most of our recycling choices. Plastic is made from gas and oil, so when oil prices are low, so is the price of new plastic. Not only is recycled plastic more expensive, its quality is lower, so most manufacturers will choose new plastic instead.

INCINERATION

Almost 12 percent, or one eighth, of our trash was burned in incinerators in 2018. Modern incinerators burn garbage and produce electricity. Very simply, it goes like this: burning → produces heat → produces steam → drives turbine → produces electricity. Burning trash doesn't make it go away completely; you'll know that if you've ever seen a campfire after it's burned out. But the ash created by incineration weighs about 75 percent less and takes up around 90 percent less space than the trash it comes from.

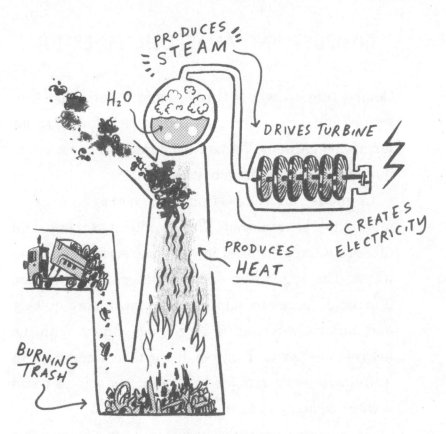

Incinerators, also known as waste-to-energy plants, are more common in Europe than in the United States. Burning trash produces enough energy to heat 1.25 million homes in Sweden, where about half of all household trash gets incinerated. One reason incineration isn't as popular in the United States is that the plants are expensive to build and maintain. Landfilling is just cheaper.

COMPOST AND ANAEROBIC DIGESTION

Composting is a way to turn organic material like food scraps and grass clippings into a rich fertilizer for farms and gardens. We sent 8.5 percent of our trash to industrial composting facilities in 2018.

In industrial composting, any nonorganic items must first be removed. Things like packaging and utensils often get mixed in with food scraps and plate waste. The remaining organic material is shredded into small pieces to help it break down more quickly and uniformly. It's put in large piles, where, with the help of moisture and oxygen, it starts to decompose. In a well-managed compost pile, your food scraps could become compost in forty-five days!

Organic waste can also be put into something called an anaerobic digester. Basically, a digester is a large sealed tank with some really cool microbes in it that munch on carbon-based feedstock like food scraps and manure and turn it into methane. Methane is one of the gases produced by landfills, but anaerobic digesters produce and capture it more efficiently. Methane from digesters can be used as gas or turned into electricity.

Like incinerators, anaerobic digesters are more

common in Europe than in the United States. Some U.S. wastewater treatment plants are experimenting with adding food scraps to the digesters they use to process sewage. They're also catching on with some dairy farmers as a way to handle all that manure. Those microbes must be pooped!

MISMANAGED WASTE

We've seen most of the ways we handle our trash, but some waste never makes it to a proper disposal site. Street litter is an obvious example of that. What's less obvious is the trash that washes into waterways and eventually makes it to the ocean. In 2016, over 2 million tons of plastic was classified as mismanaged in the United States. That's plastic that either ended up as litter, was illegally dumped, or was sent to another country to be processed, as we saw earlier in this chapter.

In 2010, the world let almost 9 million tons of plastic get into the ocean from our coastlines. If we don't change our habits, by 2025 it's estimated that we'll be dumping over 19 million tons of plastics per year into the marine environment. And this isn't even counting things like plastic fishing nets or other equipment from aquaculture or plastic that washes into the oceans after natural disasters like tsunamis.

Globally, we mismanage about a third of all waste—that's over 700 million tons of garbage being handled in ways that aren't safe for the environment, or for us.

GROSSLY MISMANAGED WASTE

Speaking of garbage in the wrong place, your plumbing is only meant to handle poop, pee, and toilet paper. Ignoring this fact can lead to truly terrifying consequences.

Smelly, bacteria-laden globs of cooking grease, baby wipes, and personal hygiene products, known as fatbergs, are developing in sewers in the United Kingdom and the United States. Instead of floating down to the wastewater treatment plant to be processed, the wipes and congealed grease, along with tampons and other things that might be too gross to mention here, stick together like gum and the sole of your sneaker, attracting all the grease and wipes that come after them.

They can grow to unbelievable sizes. The longest of these putrid monstrosities was over eight hundred feet long and weighed about 143 tons. That's like having 1,146,080 sticks of rancid butter clogging up a sewer pipe. Get your popcorn ready.

What happens if a fatberg stays in the sewer? If it's not discovered and removed in time, an enormous fatberg could block the sewer completely. Everything that you sent down your toilet wouldn't be able to get past the fatberg, and it would all come back out of your toilet, spilling over your bathroom floor.

It's a good thing some people have the enviable job of getting fatbergs out of sewers, isn't it? Fatbergs have to be broken down chunk by revolting chunk using high-pressure hoses, scrapers, and vacuum

systems. The stench is overwhelming even for workers who are used to spending their days amongst all the usual inhabitants of the sewer.

But just because they're disgusting doesn't mean fatbergs can't also educate and inspire. A slice of London's most famous fatberg was even put on display at the Museum of London!

A HUGE WASTE

One final issue with all of these methods of waste disposal: They don't do anything to encourage consumers or manufacturers to make less trash in the first place! You could even argue that waste-to-energy and recycling programs make us feel less concerned about the amount of garbage we create. We've been taught that sending things to the landfill is the most wasteful thing we can do, so anything must be better than that.

Recycling has become a particularly useful tool for the packaging industry and plastics manufacturers. When communities said no to incinerators and landfills, the packaging industry promoted recycling as an answer. It allowed consumers to feel better about consuming, which let manufacturers keep manufacturing.

You can see that, when it comes to your trash, there are no easy answers. Every solution comes with problems of its own. Are you starting to feel like your trash just keeps piling up, like some kind of immense garbage monster that threatens to consume you at every turn?

Well, maybe that garbage monster is just trying to tell you something.

Chapter Five

WHAT CAN WE LEARN FROM OUR GARBAGE?

We live in a world full of stuff. We've already seen that our stuff, including our trash, never really goes away. You're probably thinking about your trash more than you were when you started this book, but have you thought about what your trash can teach you?

ART

Among the guests at the Paris Climate Conference in 2015, where world leaders met to discuss climate change and make global agreements about carbon emissions, was an unusual creature that reminded everyone of how serious their job really was. Climatesaurus, a long, snaking dinosaur made of hundreds of thousands of plastic bottle caps, wound through the streets and climbed the steps leading to the conference.

Climatesaurus was the creation of Italian artist Maria Cristina Finucci. Plastic, of course, is made from oil, which comes from decomposed prehistoric plants and animals. You could hardly find a more powerful symbol of a crisis that began with our overuse of fossil fuels than a dinosaur made from dino-fuel.

Artists like Finucci use garbage to make us think seriously about garbage pollution and its effect on the environment, and our habits of consumption. It might seem surprising, but our trash can teach us a lot about ourselves and our world.

Climatesaurus isn't Finucci's only artistic statement about the environment. She's also the founder of the Garbage Patch State, an imaginary nation named for the Great Pacific Garbage Patch. The Garbage Patch State exists online and in a series of art installations made from trash known as the Wasteland cycle. Finucci has even planted a Garbage Patch State flag and opened an embassy in an Italian museum to raise awareness of the plastic crisis. You might not be able to book a flight to the Garbage Patch State, but the project reminds us that the threat of marine plastic pollution is very real.

On the other side of the globe, artists Richard Lang and Judith Selby Lang have collected plastic that's washed ashore along a stretch of beach at Point Reyes National Seashore in California for more than twenty years. Their ongoing project, called One Beach Plastic, is a series of arrangements of plastic items they've picked up over the years, often placed together by color or shape. They regularly "shop" on the same 3,000-foot section of the coast, looking for certain types or colors of beach plastic. A two-hour trip can end with the pair hauling seventy-five pounds of plastic back home! One of their finds shows us how long plastic can stay in the environment without breaking down: a toy oil truck from the 1940s.

Another art project created from ocean plastic is called Washed Ashore. Based in Oregon, Angela Haseltine Pozzi gathers found plastic and, with a team of volunteers, creates large, detailed sculptures of marine animals—a parrotfish, a sea lion, an octopus, a great white shark, and more. Each sculpture is made using a metal frame, and the pieces of plastic, carefully sorted by color, are attached by wire or screws. It can take years to collect enough pieces of plastic of the right color for a single sculpture.

But these animals are tiny compared to a creature made from plastic found on a Hawaiian beach. Skyscraper the whale, created by designers Lesley Chang and Jason Klimoski, stands nearly four stories tall and the plastic alone weighs five tons! The team collected plastic for four months, then assembled everything they found, including laundry baskets, fans, trash can lids, barrels, baby bathtubs, bottles, and more on a giant steel frame. The whale was placed in a canal in Bruges, Belgium, as part of a city-wide art event in 2018.

Each of these artists hopes to educate and inspire people to do something about the plastic problem, but turning trash into art can make a different kind of political statement too. Along a beach preserve in Mexico, artist Alejandro Durán installed a series of works on site, made with the waste he found along the shore. Colorful bottle caps, fields of dirty toothbrushes, and rivers of single-color plastics from over fifty nations contrast against the natural beauty of their setting. Durán's work isn't just about the environment; it's also about inequality and power. The garbage of rich nations often becomes a problem for the poorer nations that end up living with its consequences.

John Outterbridge, a Black artist in Los Angeles, used castoffs from his South Central neighborhood to create an artistic statement about what it means to be a marginalized person in America. Using found objects (an artistic term for stuff that's been thrown away) was a way of showing that what one group rejects, another can embrace and make into art. His sculptures used rags, hair, metal, wood, and old painted canvases to reflect Black history and culture.

As a child, Outterbridge heard Chicago rag men calling out to collect old fabric scraps, so rags and old clothing became an important part of his art. Other artists have gone in the opposite direction and used all types of trash to create clothing! Trash fashion, or trashion, can be jackets, hats, jewelry, or even elaborate gowns made of everything from beach trash to cardboard and newspaper. These dresses aren't made to be worn every day. They might be created specially for an art exhibit or trashion runway show to help spread the word about sustainability.

Have you heard the saying *You are what you eat*? What about *You are what you throw away*? Okay, that's not really a saying, but British artists Tim Noble and Sue Webster might say it should be. Their unique take on turning garbage into art is to collect months' worth of their own trash and found objects from their London neighborhood and painstakingly assemble it into a sculpture that looks like a normal trash pile on first sight. The magic happens when a light is placed at a particular angle. The silhouettes cast against the wall are revealed to be Noble and Webster themselves in various poses, as if to say, *Surprise!*

You can decide for yourself just what their message is (that's the great thing about all art), but it's hard to deny that literally building an image of yourself from your own trash is a pretty powerful statement about our throwaway society.

ARCHAEOLOGY

When you think of archaeology, what's the first image that comes to mind? A pyramid? A legion of terra-cotta warriors? These impressive remains of past cultures give us a sense of the grandeur of their leaders' lives, but what about ordinary people? That's where trash comes in.

You could almost say that everything archaeologists study is trash. But that's not exactly true. Just as not every pile of beach plastic is a sculpture, not every object buried under centuries of dirt and discovered by an archaeologist is trash. So how can they tell the difference?

Like detectives, archaeologists rely on context clues to understand the objects they find. Was the object found inside or outside? Was it whole or broken? What other objects were found with it? A broken object found outside with several other broken objects is likely to be garbage. An object found within the walls of a building was probably still being used.

We saw from the history of garbage that societies in the past didn't treat their trash the way we do today. For them, *away* might have been no farther than their backyard, in an old well or pit. Or they might have kept it even closer to home, literally building it into the walls of new construction projects. In Çatalhöyük, Turkey, archaeologists have even found human and animal poop between the walls of buildings, which piled up over time and became a foundation for new homes! Street levels rose over time in many places too as the garbage of previous generations became the ground beneath the next generation's feet.

Middens, or archaeological garbage heaps, don't have to be ancient to be interesting. At a prisoner of war camp in Manitoba, Canada, where German soldiers were held during World War II, archaeologists studied broken cups and bowls in the camp's dump to learn about the way prisoners were treated. The Geneva Convention states that POWs should be treated by the same standards as a country's own soldiers, and the remains of the mess hall dishes suggest that conditions in the camp were sanitary and that the prisoners were given enough to eat.

Studying trash with the tools of an archaeologist can also lead to a better understanding of social justice issues like homelessness, also known as houselessness. In surveys of things left behind at abandoned encampments, researchers learned that many of the things that well-meaning people gave to the unhoused weren't always useful to them.

Travel-size personal care products were often left behind, as unhoused people had nowhere to use them and didn't lead the kind of stable lives that let them brush their teeth or shower regularly. Secondhand shoes might also be discarded: While they seem valuable, they might be too worn or uncomfortable to be useful, or they might be the wrong size. Seeing the objects abandoned as garbage by unhoused people gives us a better sense of how we can actually help someone in that situation instead of relying on our own socially biased ideas.

We took a quick look earlier at nineteenth-century mudlarks, children who hunted along the Thames riverbanks in London for anything they could sell for a few pennies. Today's mudlarks, mainly adults, are sometimes treasure hunters but often have an interest in amateur archaeology as well. Unlike most rivers, the Thames has tides, and at low tide, its muddy banks are revealed along with the treasures of several thousand years of history. Mudlarks search the river's shore looking for everything from Roman coins to scraps from pieces of armor made at the Tower of London.

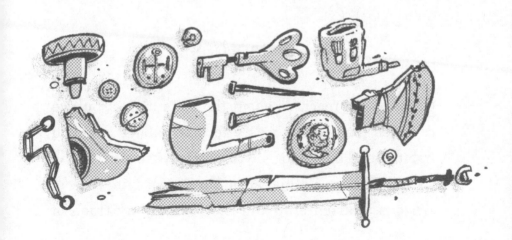

Not just anyone can pick things out of the Thames mud today. To be on the shore mudlarking at all, you need a permit, which lets you dig down nearly three inches. To dig deeper than that, up to four feet, you need a special permit that's only given to members of the Society of Mudlarks. Often, mudlarks bring their finds to experts for identification and sometimes even donate them to museums if they're rare enough.

Victorian London's sewers had their own special scavengers, known as toshers, who prowled through sewage, looking for coins and silver that washed down the drains. They carried lamps and long sticks to defend themselves from the rats that made the sewers their home.

GARBOLOGY

Garbage can teach us about the distant past, but we can also look to our waste to learn about our own habits—and bust a few myths while we're at it. In the 1970s, William Rathje, an anthropologist in Tucson, Arizona, started studying the present-day trash of ordinary people. (His group found the twenty-five-year-old guacamole!) With the help of sanitation workers, Rathje excavated landfills using a bucket auger, a drilling machine that can dig up to 110 feet into a landfill and pull out a cylinder of trash. By looking for newspapers or other items that had a printed date, the team could determine how old each sample was.

The team also collected curbside trash. They found that when they studied the households' garbage and then asked them to fill out a survey, people weren't always right about their own habits. They tended to say that they drank less alcohol and ate less processed food than they actually did and that they ate more vegetables. Maybe they didn't remember, or maybe it was the effect of something called survey bias, which makes people present themselves in the best way possible when their answers are being recorded—and possibly judged.

Studying modern garbage also helps us to understand how we produce it. For example, Rathje found that when curbside bins got larger, people threw away more stuff. The amount of trash we make expands to fill the space we have available!

Another surprising finding: There was always more

hazardous waste in the regular waste stream after a special hazardous waste collection day. Why? People had been collecting their batteries, oil cans, and other toxic waste, but if they didn't make it to the drop-off site on the single day it was open, they just threw it all in their curbside bins instead of hanging onto it until the next hazmat collection day.

Some of our supposedly good shopping habits lead to more waste as well. Buying in large bulk or family-size packages can help to save money and reduce waste, but if our bulk foods go bad before we can finish them, we're wasting money and creating trash. The same goes for stocking up on sale items—if we buy more than we can use and end up throwing food away, it's wasteful in all senses of the word.

You're also more likely to waste food if you're always buying different kinds of foods. The specialty vegetable or new type of cereal might have seemed appealing in the store, but in our busy lives, we're more likely to cook a few staple dishes again and again. It might sound boring, but sticking to the things you know you'll eat—and that you know how to cook—is a better way to avoid food waste.

Digging through dumps helped Rathje see that our trash looks very different from how we imagine it.

Forty years ago, he pointed out that the most common type of landfill trash by far is paper, which can almost always be recycled if it's not contaminated. Unfortunately, we still landfill almost a third of our paper waste. We *can* learn from our garbage, but whether or not we actually do is another question.

ON THE TRASH TRAIL

We looked at where our garbage goes: landfills, recycling centers, incinerators, and so on. But do you know *exactly* where your trash goes? Which recycling facility did your bottle go to? How far away from you is it? And do the environmental benefits of recycling the bottle outweigh the environmental costs of getting it there?

A study by the Senseable City Lab at the Massachusetts Institute of Technology in 2009 invited five hundred people to attach small tracking devices (the guts from old flip-style cell phones) to three thousand different pieces of trash, from banana peels and sneakers to lithium-ion batteries and printers. The team then traced the signals to see where each item went.

The longest journey of any piece of trash was taken by a printer cartridge, which traveled over 3,800 miles! That's a lot of carbon emissions, more than would have been generated by tossing the cartridge in a landfill. It's still important to recycle e-waste—remember the problem of toxic chemicals leaching from the landfills? Overall, e-waste and hazardous waste travel the farthest, because they're handled by specialized facilities that aren't as common as the recycling centers that handle materials like paper and glass.

The MIT researchers also learned that two pieces of nearly identical trash (two old cell phones, for example) can start out in the same city and travel very different routes to end up at the same e-waste recycling facility hundreds of miles away, creating more carbon emissions than necessary. By studying the trail of real trash, we can learn the most efficient way to get it to its final destination.

Metals took some of the shortest trips of all materials, just another point on the sustainability scoreboard for our old friend the aluminum can.

IT'S NOT JUST YOUR GARBAGE ANYMORE

Garbage has even made it all the way to the Supreme Court of the United States! A man in California was arrested on drug charges based on a warrant that a police officer obtained after going through his trash. His lawyers argued that a person's garbage was their property, even when it was in a curbside bin waiting to be picked up or in the hands of a sanitation worker. That meant it was protected by the Fourth Amendment to the Constitution, which prohibits "unreasonable" searches of personal property.

However, the Supreme Court ruled that once a

person has brought their trash out to be collected, they can't expect it to remain private. After all, they're giving it away. It's not their property anymore.

Your trash might get you in trouble, or it might teach you about art, archaeology, or environmentalism. Whatever the case, garbage has something to say. What do you think your garbage says about you?

Chapter Six

HOW BAD IS OUR GARBAGE PROBLEM?

Why is garbage a big deal? Litter is ugly, and we don't want marine animals to eat plastic, but isn't it okay to make and use lots of stuff if we throw it away responsibly?

In this chapter, we'll see that our garbage isn't just our problem. It's a problem for the environment, human health, and social justice.

GARBAGE AND CLIMATE CHANGE

We saw that organic waste produces gases, mostly methane and carbon dioxide, as it breaks down. These are both greenhouse gases, meaning that when they get into our atmosphere, they trap heat, making Earth warmer and contributing to other effects of climate change.

However we dispose of our food scraps and other organic waste, some amount of methane and carbon dioxide will be produced. There's no way to avoid creating some amount of carbon-based gas, because organic waste, by definition, contains carbon. That's just science. But *how* we dispose of organic waste makes a difference.

Let's put our chemistry goggles on for a moment.

When organic waste breaks down in the presence of oxygen, it produces carbon dioxide. When organic waste breaks down without any oxygen present, it produces methane. Got it so far? Although methane doesn't last as long in the atmosphere, it's about twenty-eight times more powerful at trapping heat than carbon dioxide. That's the carbon dioxide equivalent, or CO_2e. The best way to handle our food scraps is the way that has the lowest CO_2e.

Over half of the food we waste, 35 million tons, ends up in sanitary landfills. Some of the waste in landfills breaks down in the open air, which contains oxygen. That produces carbon dioxide. Some of it breaks down where air has been forced out as the compactors squeezed everything together. That produces methane. Landfill gas is about half carbon dioxide and half methane.

Industrial composting happens in large open-air piles that are turned frequently to bring more oxygen in. Composting creates mostly carbon dioxide and very little methane. Anaerobic digestion creates only methane because the digester tank is sealed, so no oxygen can get in.

Simple, right? With this information, it looks like composting is better than landfills, and landfills are better than anaerobic digesters.

Not exactly. Now that we've made all that gas, we need to see what happens to it.

In modern landfills, gas can be captured and turned into electricity or used for heating or other energy needs. But, because the landfill cells are open as they're being filled, gas can also escape straight into the atmosphere. Once a cell gets covered, it's easier to collect the gas, but so much has already left that a landfill might only be able to capture about 20 percent of the total gas produced over its lifetime.

Landfills can also choose to simply flare the gas—that is, to collect and burn it off—which turns any methane into carbon dioxide. It's better than putting methane into the atmosphere, but it's still pretty wasteful.

Compare that to what happens in industrial composting facilities. They don't capture the carbon dioxide they produce either, but they do create something very useful—compost. Adding compost to soil not only improves the soil, it means that farmers can use less synthetic fertilizer to grow their crops. Double win!

Anaerobic digesters produce only methane, but because the tank is completely enclosed, all of that gas can be captured and used. The leftover material in the tank, known as digestate, can also be used as fertilizer or bedding for farm animals. So why aren't anaerobic digesters more common? They require far more resources to build and they're much more expensive than landfills or compost facilities—yet another example of how our waste problems often come down to money.

This is a simple way to look at the differences between food waste pathways. Keep in mind that there are other factors too, like the fuel used to power all the equipment that runs each of these facilities, and even the trucks that deliver the waste.

If all those things were equal though, composting

all the food we sent to landfills in 2018 would be like taking nearly 11 million cars off the road!

Quick question: What's better than capturing gas or creating compost from millions of tons of food waste?

If you said, *wasting less food*, give yourself an imaginary gold star!

GARBAGE AND GROUNDWATER POLLUTION

The stuff that leaks out of landfills is another serious problem. As rain filters through the waste, toxins can leach out of garbage and down into the soil and groundwater. Although the EPA requires modern landfills to collect and treat their leachate to remove contaminants, some older landfills weren't built with collection systems.

As we use more plastics and synthetic chemicals, more of our garbage has toxic potential. A family of chemicals known as PFAS (Per- and Polyfluoroalkyl Substances) have been found in landfill leachate. PFAS come from cookware with nonstick surfaces, stain-resistant furniture and carpeting, some types of food packaging, and adhesives. Researchers are still studying their effects on human health, but some studies show that PFAS can cause problems with human growth and development, reproduction, hormones, and our immune systems.

Then there's household hazardous waste. Batteries and electronics, LEDs and fluorescent lightbulbs, cleaning chemicals, weed killers, motor oil, and paint are all banned from regular landfills, but that doesn't stop people from putting them in the trash anyway.

Most landfills only have to keep their monitoring systems in place for thirty years after a landfill is closed. That might seem like a long time, but landfill pollution can last much longer than that.

BURNING ISSUES

Incineration creates energy and reduces the volume of our trash. So why aren't we using it everywhere? There are a few problems with incineration, starting with what lurks in the ash it creates. Burning trash, whether it's in an industrial incinerator or a backyard burn pile, creates chemical compounds called dioxins.

There are many types of dioxins, and some cause cancer.

Dioxins in incinerator smoke rise into the atmosphere and then fall to Earth, contaminating soils and entering the food chain. Dioxins have been found in animal feed, and because they can be stored in fatty tissue, they build up in animals raised for food. Dioxins enter our bodies if we eat those animals.

Incinerators have systems called "scrubbers" that can remove most, but not all, of the dioxins and other toxins from the smoke that vents into the atmosphere. Wet scrubbers use water and other liquids to remove toxins, and dry scrubbers use other chemicals to do the job. Some incinerators use both types. But, just as there's no *away* for your household garbage, there's no *away* for dioxins either. The scrubber traps the chemicals, but they still pose a health hazard. Because they've trapped the toxins that would have gone into the air, both wet and dry scrubbers create their own toxic waste. The scrubbers have to be taken out and disposed of, giving the dioxins another chance at getting into the environment.

The ash produced by incineration also contains dioxins and other toxic substances that can get into the soil and water when the ash is buried. In some cases, incinerator ash is used in agriculture or manufactured into construction materials and used to build and repair roads. Dioxins that get into the soil from ash can end up in the food chain as well.

Like landfills, incinerators claim to be green energy sources. The truth is that it's impossible to produce as much energy from burning trash, or capturing the gas it creates, as it takes to make the trash in the first place. More science! This is also true for the gas created by landfills and anaerobic digesters. That's why making less garbage is always the best choice for the environment.

Even recycling is more energy-efficient than incineration. Because of all the upstream costs of paper manufacturing, for example, recycling a ton of paper saves over four times the amount of energy that could be produced by burning a ton of paper!

PLASTIC, PLASTIC EVERYWHERE

Plastic waste has literal downstream effects as it washes into our waterways and oceans. Unlike the animal carcasses dumped by nineteenth-century sanitation workers, plastic doesn't biodegrade. It photodegrades, meaning that over time, sunlight breaks it into smaller and smaller pieces. The smallest of these are called microplastics. Some plastic pieces look like food to fish, marine mammals, and seabirds, who eat them and then starve because their stomachs are full of plastic instead of food.

Large pieces of plastic have been found inside dead marine animals too. A plastic shopping bag floating in the water can look a lot like a tasty squid. Over 100,000 marine mammals are killed each year by eating plastic, including whales, dolphins, and seals. Even more seabirds die from filling their stomachs with plastic—over a million each year.

The Great Pacific Garbage Patch that inspired Maria Cristina Finucci's Garbage Patch State isn't just a pile of trash floating on the surface of the ocean. It's also not the only ocean garbage "patch." Ocean currents across the globe create five different gyres, which are sort of like whirlpools that collect debris in the water. Because many of the pieces have broken down into microplastics, you could be in the middle of a garbage patch and not see any garbage at all. In fact, surface plastic might be as little as 1 percent of all the plastic in the oceans. The rest is beneath the surface. Over 15 million tons of plastic is resting on the ocean floor. That's the equivalent of 570 billion plastic twenty-ounce bottles!

What's the most common type of ocean trash? Cigarette butts. The filters in cigarette butts are made of plastic. Filters also contain toxic chemicals like lead, formaldehyde (the stuff insect collectors use to kill their specimens), and arsenic, a kind of poison! Coastal cleanups in 2017 reported finding almost 2.5 million cigarette butts on beaches around the world!

When we talk about the marine plastic crisis, we can't leave out a sneaky type of waste that gets into the oceans directly through your pipes. It's not fat-bergs this time: Plastic microfibers wash off your synthetic clothing and travel through pipes and sewers to wherever your wastewater gets discharged.

We saw that some plastics are downcycled into clothing. In one study, a fleece jacket was found to release over 81,000 plastic microfibers in a single wash. That might be hard to picture, but consider this: The same study compared washing 100,000 jackets to putting nearly 12,000 single-use plastic grocery bags directly into the oceans. Altogether, we send over 550,000 tons of these microfibers into the ocean every year.

If we don't do something about this now, by 2050 the weight of all the plastic in the ocean could be larger than the weight of all the fish.

Microplastics aren't just found in the oceans. They're everywhere, from the soil in Antarctica to the snow and streams of Mount Everest, the highest peak in the world. These are probably left behind by modern climbers, whose technical gear is often made of nylon, polyester, and acrylic.

It's obvious that mismanaged plastic waste is a problem. But what about simply recycling more of our plastic or making sure that none of it gets into the oceans and soils?

Remember that we've only managed to recycle 9 percent of all plastics ever made. There's room for improvement, but plastics recycling is confusing and not always profitable, due to the many types of plastics out there. The more times we recycle plastic fibers, the lower the quality gets, so all plastics will have to be downcycled or ditched eventually.

Even if we could recycle all of our plastic, we have to consider the upstream problems too, beginning with drilling for oil and gas. Many communities near oil fields have higher rates of cancer, respiratory disease, and other problems. The chemicals added to raw petroleum to make plastics can be toxic. If those chemicals leak or get dumped into the environment around plastics factories, more environmental and health problems happen.

Some new research suggests that plastic itself emits methane, meaning that as your old single-use water bottle photodegrades in the ocean, it might be sending greenhouse gases into the atmosphere.

Again, the best way to handle our plastic waste is to make as little of it as possible.

E-WASTED

Did you know that the most valuable part of an old electronic device might be the copper wire or other precious metals inside?

Like aluminum, copper can be melted and recycled over and over again. But it's not easy to get metals out of old electronics. Copper wires are covered in plastic insulation, and metals like gold and silver are fused to circuit boards. We can't just cut out the old wire and use it again because of safety standards. The insulation has to be removed, the copper has to be melted and re-formed, and then it can be properly recycled into something new.

The question is, who's going to do all that work? In higher-income countries like the U.S., it costs too much to pay people to do the work. We don't know exactly what happens to most of the e-waste in the world, but some of it goes overseas, probably to countries in Asia and Africa, to be recycled. The cost of labor in these countries is cheaper, which means more profits for whoever is selling the copper.

Sending e-waste overseas has another economic advantage for businesses: It's cheaper to ship e-waste to countries that don't have formal recycling infra-structure or strong environmental regulations than it would be to handle e-waste responsibly. Most of the e-waste that goes overseas is handled by informal waste workers, like the waste pickers we saw in Chap-ter 4. Because they often don't have the resources to pro-cess e-waste properly, they can suffer health problems from burning electronics to remove the precious metals and from working with toxic chemicals.

What happens if we toss our old devices in a landfill and mine new copper and other metals instead? For one thing, there's only so much precious metal in the world. They aren't renewable resources, so we don't want to waste them. And mining is an environmental disaster in its own way. It clears forests and produces sulfuric acid, which kills plants and poisons wildlife. It also uses energy, as we saw with aluminum mining.

Throwing e-waste in the landfill doesn't just mean ditching precious metals. E-waste also contains toxic metals like lead, mercury, and cadmium. These metals can cause kidney and liver problems and affect the development of human embryos. Remember landfill leachate? You definitely don't want toxic metals getting into the groundwater or the soil.

EXTERNALIZED COSTS

Again and again, trash trouble comes down to money versus the environment. It's just cheaper to pollute than it is to try to control pollution. The economic term for this is *externalized costs*. That means that part of the cost of a product isn't paid by the company that makes it or by the person who buys it. It's cheaper for the company to burn e-waste than it is to handle it properly, but the environment and the people who live in it pay the cost with air pollution and health problems.

Mass production made many of the things we use every day less expensive. This made it easier to buy more, and more buying led to more waste. Not only do some of the products made in factories use chemicals that pollute our air and water if they're not treated properly, but making more products means extracting more resources like oil, cotton, wood, and metals from the natural environment.

WHO PAYS THE COST?

In 2012, over nine pounds of landfill methane was produced for each person on the planet. This is just an average—in reality, higher-income people produce far more and lower-income people produce far less.

But we don't have one atmosphere, or one climate, for the wealthy and one for the poor. Eventually, all those emissions are going to affect everyone on Earth, whether they contributed much or not. The same goes for our oceans. Richer countries send some of their plastic waste overseas because it's easier and cheaper than trying to deal with it at home. No matter where plastic comes from, if it gets into the oceans, it becomes everyone's problem.

When it comes to the environment, the reality is that often, "everyone's problem" is worse for poorer people. If the climate changes and oceans rise, or if pollution makes one area dangerous to live in, the wealthy can move. Poorer people, who don't have as many options, will struggle. More of the industries that create pollution and handle waste are in poor communities, and these residents end up paying a higher cost in their own health. Remember the peddlers and waste pickers who collected rags, metals, and other materials that could be sold to manufacturers? With mass production, that system came to an end in the United States. In order to make huge numbers of the same thing, from plastic combs to clothing to cars, factories needed endless streams of new material, all of the same type and quality. The fastest and easiest way to do that was to use new raw materials. Using household scrap could result in products of varying quality and could even damage expensive factory machinery.

Waste pickers have always been poor people at the margins of society, trying to make a living from other people's scraps. Putting them out of business is another type of cost to society, but in countries like India and China, waste pickers actually take on externalized costs.

Informal recyclers, as they're also known, handle up to 90 percent of all recycled material in India, but in many cases, consumers and manufacturers don't pay for their labor. Instead, waste pickers generally make their money from selling the scrap they collect.

There's real economic and environmental value in the work that waste pickers in Ghana, the Philippines, and other countries do, but they often work without enough protective equipment to keep them safe from toxic chemicals in the trash they're sorting. Even though these workers are experts in the types of materials that can be recycled and they perform a valuable environmental service, their work—and their lives— aren't often valued as they should be. In 2000, a landslide of garbage at a badly maintained dump site in the Philippines killed over two hundred people who lived and worked onsite. When waste pickers and their work aren't protected, companies can keep creating waste without paying their full share of the cost.

ENVIRONMENTAL RACISM AND ENVIRONMENTAL JUSTICE

We can't talk about how our garbage problems affect the poor more than the wealthy without pointing out that racism and discrimination have forced many people of color to live below the poverty line, so poor communities are likely to have a higher percentage of people of color than wealthy communities.

The term *environmental justice* was coined by Reverend Ben Chavis, Jr. during a 1982 protest in North Carolina against a toxic waste facility in a majority Black county. Seeking environmental justice means making sure that people of color aren't paying the cost for everyone's waste problems.

Examples of environmental injustice are everywhere. The U.S. government runs a program known as Superfund to clean up contaminated sites like leaking landfills, toxic waste dumps, mines, and manufacturing plants. Seventy percent of Superfund sites are located within one mile of a public housing development, which was built to provide housing to low-income people. About half of those residents are people of color.

About 80 percent of incinerators are in low-income neighborhoods and communities with large populations of people of color. These residents have fought long and hard to keep incinerators out of their communities because of the effects on their own health and the health of their environment.

For most of the twentieth century in Houston, Texas, 80 percent of landfills were located in Black neighborhoods. It's hard to deny that environmental racism was at work in this situation, because Black residents were only 25 percent of the population.

In Chicago, a metal scrapyard in a white, wealthy part of town closed down, and the company looked to open a new scrap metal recycling plant in a poorer neighborhood whose residents were mostly people of color. Shredding scrap metal releases microparticles of metal dust, and people in the neighborhood already had concerns about the high rate of asthma.

Native nations have led the movement to protest against oil and gas pipelines. This might not seem related to trash, but remember that plastics come from petrochemicals.

Garbage workers, who often work in dangerous conditions, are especially important to the story of environmental justice and garbage. We celebrate Dr. Martin Luther King Jr.'s bus boycott and "I Have a Dream" speech, but when he was assassinated in 1968, he was in Memphis, Tennessee, in support of Black sanitation workers who were striking against unfair working conditions.

SO, IS OUR GARBAGE PROBLEM REALLY BAD?

We can confidently say that many of the effects of our garbage problem, like methane production, dioxins, groundwater contamination, and marine plastic pollution, are bad. We could also argue that it's irresponsible to encourage human beings to consume resources as quickly as we do and to consume as much as we do without thinking enough about the natural environment and our fellow humans' health and well-being.

Some of these problems seem like they should have easy solutions. But have you noticed two themes that get repeated a lot in this book?

1. There are rarely easy answers.

2. If you look hard enough, you'll find money at the root of most of our waste problems.

Maybe the best answer to our question is that our garbage problem will keep getting worse if we care more about money than the environment and the people who live in it.

Chapter Seven

HOW CAN WE DO BETTER?

We've looked at garbage through the lens of how things are: how much trash we make and the damage that it causes to people and the environment. Now we're going to look at how things could be, if we work at it.

We can't fix the garbage glut all at once. Garbage is a global problem that needs global solutions, but we can begin by thinking in levels: what can you and your family do at home, what can you do in your community, and what needs to happen in the world of corporations and governments.

YOU AND YOUR FAMILY

In 2014, photographer Peter Menzel and writer Faith D'Aluisio asked eight families to keep their trash for one week. At the end of the week, each family posed for a photograph with their trash displayed behind them. What would your family's photo look like? How much would be packaging? How much of that packaging could you avoid buying in the first place?

Here are some ways to work on making less garbage at home.

1. REDUCE AND REFUSE

You've probably already heard of the "three Rs" of garbage: reduce, reuse, recycle. The reason they come in that order is because the single most important thing you can do to create less trash is to reduce the amount of stuff, including packaging, that your family buys and uses. Reusing and recycling are also important, but now that you know a little more about recycling, you can see that it's not the best solution. We have to start by buying and using less.

Do you live near a store that sells groceries and cleaning products in bulk? When your family shops at a bulk store, you can fill your own containers and cut down on packaging waste. If there aren't bulk stores near you, try asking your family to buy larger packages—but make sure they're things you'll use! Instead of buying snack-size chip bags (that often come in another plastic bag), look for a larger bag and take a few in your lunch in a reusable container. Or can you think of a snack that doesn't come in plastic at all?

Refuse is another important R that goes along with reducing: When you're offered a straw or plastic silverware that you don't need, don't take it. There's nothing wrong with someone using a straw if they need to, as people with some kinds of disabilities do, but many people don't really need them.

2. REUSE

You don't have to live the life of a zero-waste social media influencer to make less trash. You also don't have to have fancy alternatives to single-use items.

Your great-grandparents lived the low-waste lifestyle

when they stored their leftovers in empty yogurt containers. A simple metal water bottle will last for years,
and you'll use it well over the fifty times you need
to in order to make it more climate-friendly than a
single-use plastic bottle.

Because the thin plastic film used in shopping bags
is so hard to recycle (even the collection programs at
grocery stores don't work very well), it's always a good
idea to bring reusable bags when you and your family
go to the store.

What else can you replace with durable and reusable alternatives?

3. RECYCLE

If your family doesn't already recycle, starting a collection bin is a great next step. Your trash hauler or
town transfer station can help you figure out exactly
what can go in your bin (it can be a little different in
different places), but the safe bets are:

- Newspaper, mixed paper, cereal boxes and similar packages, and corrugated cardboard

- Steel cans
- #1 and #2 plastics: water and soda bottles, milk jugs, some tubs and squeezable bottles; look for the number in the chasing arrows (recycling) symbol
- Glass
- Aluminum cans

Things that usually can't go in curbside recycling:

- Straws
- Plastic bags, plastic wrap, and plastic pillows used in shipping
- Any other numbered plastic container besides #1 and #2
- Unnumbered plastic, including blister packs (the molded, clear plastic packages that many small items like toys, tools, and pills come in)
- Juice boxes
- Flexible packaging like chip bags, juice pouches, and candy wrappers
- Mixed plastics and other things that are a mix of different materials
- Most coffee cups and lids
- Styrofoam/polystyrene

Yep, that paper cup from your last soda probably isn't recyclable because its plastic liner is too hard to separate from the paper.

Remember that even if you're allowed to mix all your recyclables in a single container, it's important to keep food waste and liquids out. These can contaminate an entire load of recycling, and contaminated recycling goes to the landfill. As much as 25 percent of all recycling is contaminated by nonrecyclable items or other waste. Rinse cans and jars, and empty all the liquids from drink containers. Remind your family to do this too!

A word about plastics: The number code on plastics, the 1–7 inside the "chasing arrows" symbol, *does not* mean that something is recyclable. The number only tells you what type of plastic it's made from. Sneaky plastics companies insist on using the chasing arrows symbol on their products because it makes everything appear to be recyclable, and we've been trained to think that if something can be recycled, it doesn't matter how much of it we use.

Putting the wrong kind of plastic in your bin can also contaminate an entire load, so make sure you're only including the types of plastic that your local waste hauler will accept.

A Note on Hazardous Waste

Never throw batteries or any kind of electronics in the trash. Your town or county probably offers hazardous waste collections days a couple times a year. It might seem like a pain to hang onto them, but regular landfills aren't built to deal with hazardous waste. You can reduce the amount of hazardous waste you make too.

4. CHOOSE SECONDHAND OR DURABLE THINGS

Does your family buy everything new? It's nice to have new things, but there's an amazing secondhand world out there. You can find fun and unique clothes, games, and decorations in secondhand stores. Thrifting can be a great way to develop your own style. Your family might even save a little money too!

Donating old clothes and toys to secondhand stores can help them get a longer useful life. But like recycling, donating isn't a guilt-free way to keep consuming more than you need. Broken things can't be resold. Neither can old clothes that are stained or torn. Some worn clothing can be made into wiping rags, but certain fabrics can't be used as wipers, and clothing with logos, designs, sequins, or glitter won't work either.

When we do buy new things, it's often easiest to buy the least expensive option. Some of us don't have much choice about that, but if you and your family do have a choice between buying something cheap that will wear out quickly or spending more for something that will last longer, really think about how you want to be spending your money. Next time you're in a big box store, look at all the plastic toys and gadgets on the shelves and ask yourself what's really worth buying. How many of them will last longer than a year? Which things will you still be interested in after a month or two?

We often think we need things that we don't. Sometimes we're influenced by commercials, or by seeing what our friends have. It can also feel good to have new things. Advertisers have encouraged that feeling, showing us images of happy people in new clothes or using new devices. That's psychological marketing. Companies aren't just telling us that their products are good, they're telling us that our lives will be better if we buy them. There's nothing wrong with wanting to be happy or even wanting to have new things. But ask yourself: Is it really the product that makes you happy? And once you have that product, how long will that happiness last? These are things to think about now and in the future, when you're making more of your own choices about how to spend your money.

5. REPAIR

Before you toss a broken appliance or ripped jeans in the trash, see if it's possible to repair it, or if you can find someone who can do the repair.

Some devices can only be repaired by the company that made them. But once you pay for something, shouldn't you be able to fix it? That idea is known as the right to repair. When we lose the right to repair something, we're stuck with a system that forces us to throw things away.

A website called iFixit makes free repair manuals available for smartphones, laptops, tablets, and even video game controllers. The site also offers information on the right to repair and other repair topics.

Library books and online tutorials are great ways to learn new repair skills as well. For non-electronics, a little paint and glue, embroidery thread, or a DIY patch can grant some extra life. And don't underestimate the power of a good cleaning!

6. COMPOST

Composting is the best way to handle food scraps, because it makes the least amount of methane and gives us something useful for our gardens. Some cities collect food scraps for municipal composting, but you can also compost at home. If your home has a yard, try making a compost pile for your food scraps and green waste. If you don't have a yard or garden, ask

your family about starting a worm bin underneath the kitchen sink. (You can order composting worms through the mail!) If you can't use the compost yourself, see if someone else needs it for their garden.

While we're talking about food, remember the ugly produce that gets rejected by supermarkets? It's not just bruised and blemished fruits that get passed over. Supermarkets often turn away carrots with extra limbs or squash that looks, well, a little squashed. If your local stores are running an "ugly produce" test to see if customers will buy less-than-perfect produce at a cheaper price, ask your family to buy some and let them know you see the beauty within these fruits and veggies.

7. MAKING CHOICES

We're constantly told that we can have it all. We can fix our problems without giving up the things we love. The reality is that you can't have it both ways. If you want to have a smaller impact on the planet, you can't live the typical consumer lifestyle. You have to make choices. You'll have to give some things up. That can be hard.

But we can't completely avoid buying things, so how can we make the best choices about what we buy? Life cycle assessments help us figure out which products contribute the least to climate change, pollution, and other effects.

You can look up life cycle assessments for some products online. However, they make some assumptions about things that might not be true in every case, like what type of fuel is used in shipping or how far a product has to travel. Plastics companies like to point out that you have to use a cotton tote bag one hundred thirty times to equal the CO_2e of a plastic bag. They're implying that it's impossible to use a cotton bag that many times, but think about it for a minute: If you go to school every day with the same backpack, you'll use it one hundred eighty times in a year! In fact, if your family goes to the store twice a week, you'll use your bags one hundred thirty times in fifteen months. When you see a figure that seems to be asking you to use plastic instead of another material, keep in mind who's making the claim and why.

Sometimes, making one good choice leads to other, poorer choices. If your family stops getting plastic bags when you check out at the grocery store, will you end up buying different plastic bags to pick up dog poop or to use in your trash can instead? How else can you handle those problems?

Take a look at the CO_2e numbers, but don't get lost or frustrated. Common sense will help you and your family make good choices most of the time.

What about biodegradable or compostable alternatives to plastic packaging and silverware? Because they're designed for a single use, they don't help us make less trash. Most "compostable" items aren't going to break down in your backyard compost pile. They have to go to an industrial composting facility, and some facilities won't accept them because they don't make good compost and they don't break down easily. Bioplastics are still plastic, meaning that they still only photodegrade, even if they come from corn instead of oil.

This is a good example of the failure of trying to have it all. No matter how sleek or exciting some eco-friendly products might be, we can't buy, recycle, or compost our way out of our garbage glut.

YOU AND YOUR COMMUNITY

Some of the things you can do in your own home are also things that you can encourage others to do in your community.

Does your school have a green team? Green teams give students a chance to learn about environmental issues and plan ways to make schools greener, from collecting food scraps for school compost piles to teaching younger students how to recycle properly and avoid food waste in the cafeteria. If you're a homeschooler or remote learner, look to other places in your community, like public libraries and after-school programs, for chances to join or start green teams.

Teachers, library staff, or leaders of other youth programs might also be able to point you toward other organizations in your community that work on environmental issues. Some kids have even started their own movements. In Flint, Michigan, teen activist Mari Copeny called national attention to the water crisis and raised money to produce water filters that she sends to communities with toxic drinking water. In the Netherlands, Lilly Platt started picking up plastic trash on her walks with her grandpa when she was seven years old. Her family helped her collect more and post pictures online. Soon, Lilly's Plastic Pickup was being followed by environmental groups all over the world.

The Repair Café movement began in the Netherlands too when environmental journalist Martine Postma realized that many of the things we throw away, like coffee makers and vacuum cleaners, could be fixed. People felt bad about throwing broken things out, but they didn't know how to fix them. At Repair Cafés, anyone can bring something in to be fixed, and handy volunteers log repair information, including common fixes and defects, into a database that other volunteers can access. Is there a Repair Café in your community? Could you help to start one?

Some of the things we've talked about in this chapter aren't available everywhere, like bulk bins at a grocery store or plastic-free produce. If you have stores that aren't offering low-waste choices, try talking or writing to the management to let them know that people in the community want those options. Let restaurant owners know that you want to see reusable take-out container programs in your town.

YOU AND THE REST OF THE WORLD

Packaging and plastics companies have encouraged us to use disposable bags, bottles, silverware, and more. Over the decades, they've used advertisements to appeal to our worries about health and germs, to convince us that convenience was more important than durability, and to sell us more and more stuff.

They've even created entire ad campaigns and charities to convince us that the resulting piles of trash created by all this stuff are our fault, not theirs. The organization Keep America Beautiful was started by two companies that made single-use cans and bottles, and its anti-litter message keeps the attention squarely on individual consumers without saying anything about the role the manufacturers play in creating waste.

Plastics companies embrace recycling programs because recycling doesn't require them to cut back on the amount of stuff they produce. Many companies are setting goals for using more recycled plastic in their bottles and other containers and claiming that this is part of their commitment to the environment.

Recycling is good, but reducing is better. Be aware of when corporations are "greenwashing" or using a pro-environmental message to cover up the way they do business. It's cheaper for corporations to support recycling than it is for them to pay for what happens to their products over their life cycle, another reason they push recycling so hard and fight other, more effective ways to deal with garbage.

Bottle bills are a simple way to increase recycling rates, but industry has fought them too. Ten states have adopted bottle bills. When you buy certain drinks (in most cases, it's beer, water, and soda, although some states also include other beverages like juice and energy drinks) in cans and bottles, you pay an extra five or ten cents per container as a deposit. When you return the containers to a redemption center or a redemption machine at a grocery store, you get your deposit back. In states with bottle bills, 60 percent of drink containers are recycled on average. Compare that to 24 percent in states without bottle bills!

Part of the reason bottle bill states have such high recycling rates is that, even if someone throws a bottle or can in the trash, a uniquely American type of waste picker known as a canner will often come along and retrieve the bottle and collect the deposit. Canning provides a living for enough people in New York City that they started their own collection center!

Instead of supporting bottle bills, drink and container

companies prefer to invest in pro-recycling marketing campaigns that are completely voluntary, and as we've seen, those are much less effective.

Similar to the fight against bottle bills, the plastics industry has fought bag and straw bans in cities and states across the country. In some cases, they've been so successful that not only did they stop a particular bag ban from being passed, they made it illegal for any city in a state to pass a bag ban of their own. Only eight states have banned single-use plastic bags in grocery stores.

Extended producer responsibility, or holding manufacturers responsible over a product's entire life cycle, is another way to make sure corporations pay more of the costs created by garbage. This idea is catching on in the United States, but in Europe it's already led to programs that charge manufacturers a fee to pay for the cost of recycling or disposing of packaging.

France now requires manufacturers of new appliances like washing machines, TVs, and smartphones to provide information about their repairability so that consumers can see how easy it will be to fix something before they buy it.

Designing products for reuse, repair, and better recycling is part of the circular economy, a model for producing goods that looks beyond the "make, take, waste" model that gave us single-use bottles and cheap fast fashion.

Pressure from consumers can show manufacturers that there's a market for repairable, reusable, and durable things. You can call, email, or write to your elected representatives to let them know how you feel about bottle bills, bag bans, and environmental regulations that prevent pollution or require companies to clean up after themselves.

PASSING THE MESSAGE ALONG

If you've read this entire book, you're probably more interested in trash than most people are. The truth is many of us simply don't think about our garbage, just as we don't all think about the environment in the same way.

We hold different values based on where we grow up, the experiences we have, and what messages we get from our family and our community. For people who are lucky enough to live in places where toxic waste, leaking landfills, and incinerators don't pollute the air and water, environmentalism is mostly about the idea of preserving the "wilderness." This is the mission of groups that raise money to save endangered species or protect natural landscapes, which are often far from where their members live. "Wilderness" in this view is a place to visit, not a place to live. There's nothing wrong with wanting to preserve nature—in fact, it's crucial—but we should remember that this isn't the only important environmental action. For people who live near pollution sources, environmentalism is a very personal commitment with immediate consequences in everyday life and in their own communities. These are both important ways to care for the planet.

Some people don't hold any strong environmental values. Some say they hold environmental values but their actions don't always match up. For some of us, choosing to buy less is enjoyable because we know we're making choices that match our values.

There is also some privilege in being able to buy certain products that are marketed as green because often those choices cost more. Not everyone has access to those products or the money to buy them.

You won't be able to convince everyone you meet that they should buy less stuff and make less trash, but it's not your job to try to make everyone else see things the way you do. Facts alone usually aren't enough to change someone's mind. Sometimes sharing facts with someone whose values are very different from your own can even have the opposite effect. That person might fight for their own beliefs even harder and resist changing because they feel like they're being judged or forced to do something or that their beliefs aren't respected.

Instead, look for the bright spots. Where can your actions have the biggest effect? Share your views and be an example to others. Offer to help people who are curious to learn more about the garbage problem. Look at the groups around you that are interested in all the issues we've looked at in this book, and put your time and energy into working with people who are already on board.

ONE LAST WORD: *HOPE*

Finally, don't give up hope.

We can't escape trash completely. We'll always throw things away, and that means we'll need places for it to go and a way for it to get there. In modern societies, garbage is part of the trade-off of convenience and choice.

Problems this big can seem overwhelming. It can feel like no solution is good enough, so why bother trying?

Don't get stuck feeling guilty about the things you buy and use. Just remember to be aware of your choices. Make the best choices you can as often as you can. Use your voice to call attention to the issues. Zero waste isn't just about making as little trash as you can personally, it's about designing a system that reduces the amount of waste everyone makes.

You can't do it alone, but you don't have to. Garbage is everyone's problem, and it's not worthless. In fact, it might just be one of the most important issues on the planet.

AUTHOR'S NOTE

My mother's parents were children of the Great Depression. They began their married life during WWII, with its ration books and "make do and mend" mentality. They never lived with serious deprivation, but like many members of their generation, they had lifelong habits of thrift. I lived with my grandparents for most of my childhood, so I learned those values too.

Open my childhood fridge, and you'd see a margarine container with the half serving of peas left over from last night's dinner. My conscientious grandmother would be eating those peas come lunchtime. My grandfather famously had a pair of woolen socks he bought in Scotland around the time I was born. He was still wearing them when I was in high school.

It's no wonder I've always been interested in stuff: what we make, what we have, what we use, what we throw away. I've tried, at various times in my life, to be low-waste, to be zero-waste (that one didn't last), to be thrifty, and to cook and mend and build things myself. I've also raised children, had jobs, driven cars, and taken vacations: In other words, I've done the sort of things that, in our modern consumer culture, are made a lot easier by the availability of plastics, packaged goods, and cheap, disposable products. After a long day's work and a long commute, I've put together a dinner of chicken tenders from the freezer, a prewashed salad in a bag, and some ready-made rolls. When my daughter tore a hole in the knee of her leggings, I bought a brand-new pair. They're only five bucks at the big box store, and it's easier than sewing on a patch.

This is all my way of telling you that I didn't write *Total Garbage* to create a template for how to live a perfect, trash-free life. Anxiety over our choices might spur us to action, but it can also make us indecisive and confused, or worse, make us give up. I wrote this book to share with my readers one of the most important things I can think of: We all live on this planet together, and we all have an impact, so let's do our best for ourselves and one another. Is that a lofty goal for a book about trash? I don't think so.

But how does a writer go about taking those two ideas—garbage and how it connects us—and turning them into a book? I wasn't completely new to the subject when I started researching this book. I'd learned about garbology, the anthropology of trash, for another book, and I'd visited a couple of anaerobic digesters for yet another project. So I had some idea of what I wanted to say and what I needed to learn. But I'm not a garbologist! I'm not any kind of scientist (unless you count library scientist, otherwise known as a librarian, and most scientists don't count that).

Librarians might not use space telescopes or mass spectrometers, but we love research. I started making plans to visit landfills and MRFs. I wanted to do a garbage truck ride-along and ask science classes to measure and categorize their trash for a week. I found a zero-waste school that I hoped to visit. I imagined that this book would get up close and personal with trash—meaning that I would get up close and personal with trash.

Then came COVID, and if you don't remember what it was like when a global pandemic was declared in March 2020, many parts of daily life came to a grinding halt. There were no kids in face-to-face classes to take on a group trash project. There was no way I was going to be allowed to sit in a garbage truck with sanitation workers. Even going to an open-air landfill seemed dangerous in a time when we were all being encouraged to stay home and avoid unnecessary travel. And to be honest, I got a little bummed out. My beautiful ideas! Now they were just, well, trash.

Then I started to notice something: Suddenly, people

online were talking about making sourdough starter. (What, natural-yeast bread baking didn't make it into your social media feeds and video recommendations? Weird.) Tons of stuck-at-home grownups were gazing at their jars of flour and water, feeding their fungi and bacteria and waiting for the magic to happen. They finally had the time to bake! And they would do more—they would start vegetable gardens, and learn handicrafts, and preserve their own tomatoes. I was one of these people. I even made my own yogurt for about a year (hey, most yogurt at the store comes in a nonrecyclable #5 container, after all).

This is, of course, a very privileged experience of the coronavirus pandemic. Many people weren't able to stay at home, and many others lost their jobs but didn't have the luxury of learning rustic skills. But the pandemic showed us a few things about our relationship to stuff:

1. We rely on convenience (and cheap, packaged goods) partly because our lives are so hectic. When we have more time, we can make different choices, like baking bread instead of buying it.
2. Rediscovering DIY culture can be fun!

It also showed us the dark side of consumerism:

1. When we couldn't spend money on experiences like movies, vacations, and eating out, we channeled our consumer energy into online shopping, sending factories into overdrive, clogging shipping channels and ports, and creating mountains of packaging.
2. We will probably never stop using disposable goods. Masks? Gloves? Swabs and syringes and rapid tests? Those are the things that kept us safe.

There's a lot to learn from the different sides of the pandemic experience. We had to wear masks, but maybe we didn't have to order all the stuff. We can't all work and go to school from home forever (and plenty of people don't want to), but maybe we can figure out how to balance our busy lives with our newfound interest in old-fashioned habits, especially ones that don't leave a trail of trash behind us.

We've always made garbage, and we always will. It just looks different at different times in history. Going through the pandemic made me see that connection in a way I hadn't before, and my limitations suddenly became my blueprint. Instead of getting up close and personal, I stepped back and took in the long view. I had so many questions about trash, I knew that readers would have them too. Hence the seven questions about waste and your world that helped me organize the information in this book.

So I put on my (reusable) rubber gloves and dug into the internet and bought a stack of books. (Yep, I ordered stuff online myself.) I looked for the weirdest, grossest, most surprising facts I could find. I looked for the problems and the solutions. I looked for the humanity in our trash.

I still read most of the articles about garbage that I come across, even though my research for this book is long over. Like all the things in your trash can, curiosity tends to stick around longer than you expect.

TRASH TIMELINE: THE GOOD AND THE GROSS IN THE HISTORY OF WASTE MANAGEMENT

7000-5000 BCE Çatalhöyük flourishes, building new roads and settlements on the trash of its previous inhabitants.

1000 BCE In Dubai, broken ceramic vessels are used as material in new tools, an early form of recycling.

500 BCE Athens outlaws disposing of trash within one mile of the city limits.

1350-1550 CE In parts of Europe, stone streets are built with gutters down the center for garbage and human waste. In England, muck carts are provided by cities for residents to haul their own trash to approved dump sites.

1500s CE Aztec leader Moctezuma II bans littering and dumping waste. In Mexico-Tenochtitlán, the largest city in the world at the time, a system of official street cleaners, scavengers, and systems for recycling food and human waste create a clean and orderly city.

1800s CE In the Ashanti kingdom in West Africa, a public works department is in charge of sanitation, including cleaning the streets and making sure residents kept their compounds clean.

1859 The first anaerobic digester is built in Bombay, India, in a leper colony.

1885 The first incinerator in the United States is built on Governor's Island in New York City.

1886 The first reduction plant in the United States opens in Buffalo, New York, where organic scraps and dead animals are boiled into a liquid used to make soap and candles.

1895 "Colonel" George E. Waring is made commissioner of the New York City Street-Cleaning Department and creates the White Wings, his corps of trash collectors and street sweepers who dress in starched white uniforms. "Source separation" begins, asking households to separate food scraps and ash from other trash.

1937 The first sanitary landfill opens in Fresno, California.

1938 The first trash garbage truck with a built-in trash compactor hits the streets.

1953 Keep America Beautiful is founded by single-use bottle and can manufacturers; their campaigns to blame individuals for litter problems take the focus off the companies' own wasteful practices.

1968   The city of Madison, Wisconsin, starts the first curbside recycling program in the United States, offering pickup of bundled newspapers. Metal recycling is added to the program in the 1970s.

1970   The Clean Air Act is signed into law, pushing landfill operators to monitor their sites for leaching toxins and leaking methane and regulating emissions from incinerators; the first Earth Day also happens on April 22, 1970.

1972   The Ocean Dumping Act prohibits dumping of "all materials" that could harm the ocean environment and human health.

1975   The PET (#1 plastic) soda bottle is introduced.

1990s   Single-stream recycling begins in some cities; households are no longer required to separate plastics, paper, and metals in their recycling bins.

1996   San Francisco, California, starts the nation's first city composting program, collecting yard waste and food scraps for industrial composting.

1997   Charles Moore, a sailor, discovers the Great Pacific Garbage Patch.

2018   China's National Sword policy begins, rejecting imports of waste plastic and paper.

2023   You take a look at the trash you create and decide to . . . ?

ACKNOWLEDGMENTS

I'm going to try to keep the garbage puns to a minimum here. Thanks to Julia Sooy for burning with enthusiasm when I told her I wanted to write about trash, and the whole team at Henry Holt Books for Young Readers/ Macmillan Children's Books for helping me make the subject easy to digest and turning my scraps into something useful and beautiful. To Molly Ker Hawn, as always, for being a reliable and renewable source of energy (and my apologies for filling your inbox with garbage). To John Hendrix for bringing his monster talents to the artwork and wasting no time in producing perfection. To the pile of experts who let me consume their time: Lori Clark of Stony Brook University; Jan de Waters, Alex French, Stefan Grimberg, all of Clarkson University; Alyssa Gunderson of Peachtown School; Nick Hamilton-Honey of Cornell Cooperative Extension of St. Lawrence County; and everyone who had to hear about "the garbage book." To Sandhya Ganapathy for letting me endlessly recycle my ideas for this project over cups of tea, mostly in mugs instead of paper cups. And to my kids, who didn't put me out by the curb every time they heard me say, "I'm not buying that. Look at all that plastic packaging!"

How'd I do?

SELECTED RESOURCES

Published for Kids

Eamer, Claire. What A Waste: Where Does Garbage Go? Annick, 2017.

Flynn, Sarah Wassner. This Book Stinks! National Geographic, 2017.

French, Jess. What A Waste! DK, 2019.

Fyvie, Erica and Bill Slavin. Trash Revolution: Breaking The Waste Cycle. Kids Can, 2018.

Kallen, Stuart. Trashing the Planet. Lerner/Twenty-First Century, 2017.

Mulder, Michelle. Trash Talk: Moving Toward A Zero-Waste World. Orca Footprint, 2015.

Published for Adults

Humes, Edward. Garbology: Our Dirty Love Affair with Trash. Avery, 2013.

Leonard, Annie. The Story of Stuff: The Impact of Over-consumption on the Planet, Our Communities, and Our Health—And How We Can Make It Better. Free Press, 2011.

Maiklem, Lara. Mudlarking: Lost and Found on the River Thames. Bloomsbury, 2019.

Rogers, Heather. Gone Tomorrow: The Hidden Life of Garbage. The New Press, 2005.

Royte, Elizabeth. Garbageland: On the Secret Trail of Trash. Little, Brown, 2005.

Web Resources

Break Free from Plastic breakfreefromplastic.org/

EPA Municipal Solid Waste Data epa.gov/facts-and-figures
-about-materials-waste-and-recycling

MIT Senseable City Lab Trash Track senseable.mit.edu/
trashtrack/

Red Ted Art upcycled craft videos youtube.com/playlist?
list=PL2vt_TPKQbZpUYcEZ6YKCGSBw_HaNL1Yc

SciShow Kids video: Make the Most of Compost! youtube
.com/watch?v=Q5s4n9r-JGU

The Story of Stuff storyofstuff.org/

INDEX